D1619783

Genserik L.L. Reniers

Multi-Plant Safety and Security Management in the Chemical and Process Industries

Related Titles

Nemerow, N. L., Agardy, F. J., Salvato, J. A.

Environmental Engineering

Environmental Health and Safety for Municipal Infrastructure, Land Use and Planning, and Industry

2009

ISBN: 978-0-470-08305-5

Center for Chemical Process Safety (CCPS)

Guidelines for Hazard Evaluation Procedures

2008

ISBN: 978-0-471-97815-2

Stoessel, F.

Thermal Safety of Chemical Processes

Risk Assessment and Process Design

2008

ISBN: 978-3-527-31712-7

Bender, H. F., Eisenbarth, P.

Hazardous Chemicals

Control and Regulation in the European Market

2007

ISBN: 978-3-527-31541-3

Bhagwati, K.

Managing Safety

A Guide for Executives

2006

ISBN: 978-3-527-31583-3

Vogel, G. H.

Process Development

From the Initial Idea to the Chemical Production Plant

2005

ISBN: 978-3-527-31089-0

Genserik L.L. Reniers

Multi-Plant Safety and Security Management in the Chemical and Process Industries

WILEY-VCH Verlag GmbH & Co. KGaA

The Author

Prof. Genserik L.L. Reniers
Universiteit Antwerpen
City Campus, Office B-434
Prinsstraat 13
2000 Antwerpen
Belgium

Library of Congress Card No.: applied for

British Library Cataloguing-in-Publication Data
A catalogue record for this book is available from the British Library.

Bibliographic information published by the Deutsche Nationalbibliothek
The Deutsche Nationalbibliothek lists this publication in the Deutsche Nationalbibliografie; detailed bibliographic data are available on the Internet at <http://dnb.d-nb.de>.

Composition Toppan Best-set Premedia Limited, Hong Kong
Printing and Bookbinding T.J. International Ltd., Padstow
Cover Design Schulz Grafik-Design, Fußgönnheim

Printed in Great Britain
Printed on acid-free paper

ISBN: 978-3-527-32551-1

I would like to dedicate this book to my wife Carmen and my daughter Kari who certainly make life worth living.

Multi-Plant Safety and Security Management in the Chemical and Process Industries. G.L.L. Reniers

ISBN: 978-3-527-32551-1

Contents

Multi-Plant Safety and Security Management in the Chemical and Process Industries. G.L.L. Reniers

ISBN: 978-3-527-32551-1

Preface

Compared to the first decades of the previous century, the number of plants handling hazardous chemicals has increased as a direct consequence of an ever-increasing variety of products and processes. Simultaneously, plants have come to be located closer to populous neighborhoods due to rising density of populations. As a direct result, the likelihood and the catastrophic impact of one accident in one chemical company causing a secondary accident in a nearby company, in turn possibly triggering a tertiary accident in another chemical plant, and so on (so-called external domino effects), increases. It is only over the last decade that politicians and industry policy makers have become more aware of the necessity to explicitly manage these very low frequency risks, making the research of external domino effects assessment and external domino effects management a priority topic. However, to date there is no existing management strategy for countering such cross-company major hazards.

This book develops recommendations and guidelines for elaborating a multi-plant safety and security culture. Such a cross-company culture has the potential to become an essential component of managing chemical safety and security in industrial areas worldwide. Besides demonstrating the use of the different parts proposed for composing such a culture, the book also focuses attention on the need to devote greater effort to cross-company disaster management.

I sincerely hope that the book will increase the reader's knowledge about managing safety and security prevention issues within and between chemical plants and thus generate multi-plant safety consciousness and multi-plant security awareness among all stakeholders. The readership who might not be fully convinced of the importance of safety and security management in a multi-plant context might like to keep the following witticism in mind:

> If multi-plant safety precautions seem to be expensive or multi-plant security requirements appear to be unrealistic, check out an external domino effect or a terrorist suicide attack. You may be truly surprised.

Genserik Reniers

Multi-Plant Safety and Security Management in the Chemical and Process Industries. G.L.L. Reniers

ISBN: 978-3-527-32551-1

List of Acronyms

ACC	American Chemistry Council
AIChE	American Institute of Chemical Engineers (US)
ALARP	As Low As Reasonably Practicable
ASCinC	Advancing and Stimulating Cooperation in the Chemical industry
ASIS	American Society for Industrial Security
BLEVE	Boiling Liquid Expanding Vapor Explosion
BPCS	Basic Process Control System
CCPS	Center for Chemical Process Safety (US)
COMAH	Control Of Major Accident Hazards (UK)
CQ	Control Question
CR	Control Room
DBT	Design Basis Threat
DDF	Domino Distance Factor
DDPF	Domino Danger Path Factor
DDSF	Domino Danger Segment Factor
DDU	Domino Danger Unit
De	Domino event
DEA	Domino Effect Analysis
EDP	External Domino Prevention
EPA	Environmental Protection Agency
ES	Electronic Source
ETA	Event Tree Analysis
FAR	Fatal Accident Rate
FMEA	Failure Modes and Effects Analysis
FTA	Fault Tree Analysis
HAZOP	Hazard And Operability (Study)
Hazwim	Hazop, What-if Analysis and the Risk Matrix (Framework)
HR	Human Resources
HSE	Health and Safety Executive (UK)
I	Invest (strategy)
IAEA	International Atomic Energy Agency
IDE	Instrument Domino Effects
IESLA	Instrument for Existing Staffing Levels Assessment

Multi-Plant Safety and Security Management in the Chemical and Process Industries. G.L.L. Reniers

ISBN: 978-3-527-32551-1

InSec	Indoor Security (Framework)
IPL	Independent Protection Layer
ISO	International Standardization Organization
JSMS	Joint Safety Management System
KM	Knowledge Management
LOC	Loss Of Containment
LTI	Lost Time Injury
MCSL	Methodology to evaluate Changes of existing Staffing Levels
MICADO	Méthode pour l'Identification et la CAractérisation des effets Dominos
MPC	Multi-Plant Council
MPDE Game	Multi-Plant Domino Effect Game
M-PSMP	Multi-Plant Security Management Program
M-PSMS	Multi-Plant Safety Management System
M-PSSC	Multi-Plant Safety and Security Culture
NI	Not Invest (strategy)
OECD	Organization of Economic Cooperation and Development
OHSAS	Occupational Health and Safety Management System
OSHA	Occupational Safety and Health Administration
OSP	Operator Security Plan
OutSec	Outdoor Security (Network)
PDCA	Plan, Do, Check, Act
PFD	Probability to Fail on Demand
PSMP	Plant Security Management Program
PSMS	Plant Safety Management System
QRA	Quantitative Risk Assessment
R&D	Research and Development
RAMCAP	Risk Analysis and Management for Critical Asset Protection
S&S	Safety and Security
SESQ	Safety, Environment, Security, Quality
SIF	Safety Instrumented Function
SIL	Safety Integrity Level
SL	Staffing Levels
SMP	Security Management Program
SMS	Safety Management System
SRF	Segment Risk Factor
SVA	Security Vulnerability Analysis
TD	Total Danger
TISC	Tipping Inducing Sub-Cluster
VCE	Vapor Cloud Explosion

1
Introduction

The art of making wise decisions is the hallmark of successful management and requires both pertinent information and good judgment. Safety-related decisions, in particular, have traditionally been based on hard-earned operating experience and intuition. As greater demand for improving the safety, security, health, environment, and economic aspects of facilities is placed on chemical companies' finite resources, the decision-making process becomes more difficult and the need for better information becomes more critical.

Company management now recognizes that simply reacting to accidents or attacks and then determining whether additional safety and/or security precautions are needed is no longer acceptable in that order–the potential effects of accidents and attacks are becoming increasingly catastrophic. Furthermore, today's technical, social and political environment also demands that decision makers take a more pro-active approach to safety- and security-related issues. Ever more thorough methods and strategies to gain an increased understanding of the significance of all kinds of (accidental and intentional) risks from the companies' operations are used.

Nowadays, a safety culture that includes a thorough understanding of the importance of safety management is common practice in the chemical industry. The safety culture describes the organization's overall attitude and commitment to safety, or as Williamson *et al.* (1997) put it, a safety culture comprises *the organizational responsibility for safety, the management attitudes towards safety, the management activity in responding to safety problems, safety training and promotion, the level of risk at the workplace, workers' involvement in safety, and the status of the safety officer and the safety committee.* Since September 11, 2001, the day the World Trade Center towers were destroyed by a terrorist attack in New York City, organizations, especially in the chemical industry, are aware of the need to shape a security culture as well, parallel to their existing company safety culture.

Organizational learning is a critical facet of an effective safety and security culture, and one that is common to a number of emerging models and approaches in the safety and security management field. However, several social and institutional barriers to effective organizational learning remain. Pidgeon and O'Leary (2000), for example, highlight two of these: informational difficulties and organizational politics. Cross-company learning through safety and security cooperation

Multi-Plant Safety and Security Management in the Chemical and Process Industries. G.L.L. Reniers

ISBN: 978-3-527-32551-1

is an example that satisfies both these factors. It is only by explicitly recognizing the obstacles to learning that arise in practice that one is able to advance the ideal of safe and secure (group-)organizational designs, and through this to counter the incubation of major failures and attacks.

Looking at the organizational safety and security trends within the chemical industry it is tempting to speculate about what the future holds. Remembering Reason's (1990) basic premise that *systems accidents have their primary origins in fallible decisions made by designers and high-level managerial decision makers*, companies start to realize that preventing major cross-plant accidents is the joint responsibility of safety management and security management decision makers in plants neighboring each other. The latter translates into a large number of individual plant safety and security cultures steadily evolving towards a limited number of multi-plant safety and security cultures within large chemical clusters.

A picture is slowly emerging of global operating companies that will set their own worldwide safety and security standards in combination with worldwide safety-management systems and security management programs. The growing complexity of processes and organizations, global companies with independent business units, corporate goal-setting policies with local implementation, out-sourcing and increased involvement by the public are all trends that policy makers and captains of industry have to take into account. Since the challenge will be to develop the most effective system of harmonized codes and standards, a decision-support system that can be implemented on an inter-organizational level to adequately manage cross-company hazards would undoubtedly be a valuable tool.

In this book, such a management strategy tool has been developed as part of a multi-plant safety and security culture, combining an original meta-technical framework for optimizing cross-company cooperation management with code for a technical evaluation model for ranking chemical installation items with respect to their potential danger in a complex industrial area of chemical plants. Chapter 2 discusses chemical safety risks and security risks in general, focusing on external domino effect hazards. The expected evolvement from a plant safety and security culture to a multi-plant safety and security culture is dealt with in Chapter 3, drawing conclusions regarding anticipated future developments. In Chapter 4, the engineering of multi-plant-management procedures is presented. Chapter 5 explains how inter-company cooperation can be organized by using new frameworks, originally developed for this goal in cooperation with an expert panel. Chapter 6 drafts code to elaborate a technical method for guiding the prioritizing of installation sequences that need to be pro-actively investigated using the frameworks. Chapter 7 offers a method to assess, to evaluate and to continuously optimize operational staffing levels in industrial areas. Chapter 8 integrates the building blocks and cross-plant management instruments elaborated in Chapters 4–7. The entanglement of the approaches results in a guideline document for drafting a multi-plant safety and security culture that can be developed for achieving optimized cross-company major accident prevention within a multi-plant area being part of a (larger) chemical cluster. In Chapter 9, game-theory is employed as a mathematical technique to convince top-management of neighboring chemical plants to invest in multi-plant safety and security. Chapter 10 summarizes the work, draws overall conclusions and offers recommendations for implementation.

2
Chemical Risks in a Multi-Plant Context

2.1
Introduction

The chemical industry covers several industrial sectors, including production of chemicals and the pharmaceutical industry but also industries manufacturing, for example, paints, varnishes, soaps, detergents, *etc.* The common element in these sectors is the handling of chemicals on an industrial scale. Companies storing and transporting hazardous materials may also be considered part of this industry.

It is essential to emphasize the importance the chemical industry has for today's society. A wide variety of some 30 000 consumer products result from chemical industrial activities. The vast majority of chemicals, about 85%, is produced from a very limited number (ca. 20) of so-called base chemicals. These base chemicals are derived from raw materials (ca. 10) such as oil, salt and water. Base chemicals are converted into so-called intermediates (ca. 300), which are still relatively simple molecules, and eventually into consumer products, ranging from plastics to pharmaceuticals.

Industry in general and the chemical industry in particular, provide the necessities for our modern-day life. In fact service industries, including financial, medical and social services, are only made possible by the wealth-producing activities of production industries and the chemical industry as a part of it.

Creating wealth goes hand in hand with taking all kinds of risks. In the chemical industry, certain so-called *major risks* can cause *major accidents* with devastating effects. One possible definition of a major accident is "an accident in the processing, storage or transport of hazardous substances which has the potential to have an off-site impact in terms of at least injury to and/or evacuation of people and/or an on-site impact resulting in the death of people" (based on Wells, 1997). Thus, a major accident can cause major detriment both on and off the site to humans, equipment, property and the environment. Due to the catastrophic consequences of possible major accidents affecting the business conditions of the chemical industry, effective control of major hazard incidents is of great importance. However, despite all precautions taken, major hazard incidents have been occurring around the world for as long as hazardous materials have been processed in

Multi-Plant Safety and Security Management in the Chemical and Process Industries. G.L.L. Reniers

ISBN: 978-3-527-32551-1

relatively large quantities. As man's technological capability has grown so has the potential for man-made disasters.

Moreover, in the chemical industry, economies of scale, environmental factors, social motives and legal requirements often lead companies to "cluster". Therefore, chemical plants are most often physically located in groups and are rarely located separately. These clusters of chemical plants consist of atmospheric, cryogenic and pressurized storage tanks, large numbers of production installation equipment, and numerous pipelines for the transportation of hazardous chemicals. In and around such clusters, dangerous goods are transported in large volumes using pipelines, trucks, ships, barges and trains. Due to the rapid development of chemical technology, there is also a continuous growth of ever-more complex chemical sites and chemical clusters with more extreme and critical process conditions. These evolutions lead to an increased likelihood of major multi-plant accidents happening and a trend of extended devastation per accident in time.

2.2 Safety Risks *Versus* Security Risks

A chemical hazard can be defined as a set of circumstances that may result in harmful consequences. The probability of this resulting, combined with the severity of the damage and with the vulnerability of the assets damaged by the unwanted outcome, comprises the chemical risk associated with it. Chemical risks are thus a consequence of the presence of hazards. Such risks can be caused by either accidental circumstances, or by intentional circumstances. The former risks are termed *chemical safety risks*, while the latter are known as *chemical security risks*.

The differences between safety risks and security risks can be understood by more thorough explanation of the two types of risks. The term "risk" implies probability, not certainty.

A safety risk is defined by CCPS (2000a) as "a measure of human injury, environmental damage, or economic loss in terms of both the incident likelihood and the magnitude of the loss or injury". The definition of a safety risk thus bears the suggestion of being accidental. The incident likelihood may be either a frequency (the number of specified incidents occurring per unit of time, space, *etc.*) or a probability (the probability of a specified incident following a prior incident), depending on the circumstances. The magnitude of the loss or injury determines how the risk is described on a continuous scale from diminutive to catastrophic. As pointed out by Kirchsteiger (1998), risk is not simply a product type of function between likelihood and consequence values, but an extremely complex multi-parametric function of all circumstantial factors surrounding the event's source of occurrence.

Somebody who explicitly intends to cause damage to chemical facilities or for example, perpetrate a theft of chemicals makes for a very different *security risk analysis* than is typically conducted to assess accidental safety risks. A security risk (suggesting intentionality) is an expression of "*the likelihood that a defined threat will exploit a specific vulnerability of a particular attractive target or combination of targets to cause a given set of consequences*" (CCPS, 2003).

In general, risk is understood as being a broad concept with no clear classifications; in other words, to date different categories of risk have as such not been defined. The term *chemical risk* is used to indicate that the risk is caused by the presence of chemical substances.

It should, however, be stressed that certain chemical risks, safety-related as well as security-related, have the potential of causing off-site detriments and losses. Hence, chemical plants cannot be regarded as "safety islands" in the chemical cluster where they are situated. Most companies are indeed – through safety and security off-site risk ties – directly or indirectly linked with other companies composing the cluster. This book is concerned with how to optimally take preventive measures as regards risks involving more than one chemical plant (or so-called multi-plant chemical risks).

2.3
The Safety-Risk Spectrum

By studying the history of hazardous incidents in any class of engineering activity, and classifying the incidents according to the magnitude of the related loss, it may be noted that incidents with major consequences occur relatively rarely, and most common types of incidents have relatively minor effects, as illustrated in Figure 2.1. The figure is based on information concerning safety risks and safety incidents. However, Figure 2.1 is assumed to hold in the case of security risks as well, be it somewhat different (see Section 2.4, Figure 2.2).

The top left-hand end of the straight line in Figure 2.1 encompasses the field of occupational safety, where the severity of accidents is often limited by the energy available to a person in his or her workplace and may result from personal strength, hand tools, *etc.* The lower right-hand straight line end in Figure 2.1 represents the field of major hazards, or process safety, where the severity of accidents results from the energy or toxicity of the materials in process, or from the sheer scale of the operation. The fact that any particular organization may never have had a major accident paradoxically leads to a problem for management. The absence of history of accidents is no indication as to whether or not the company is likely to experience a disaster in the coming year. The lost-time injury rate in an oil refinery for example is no indication of the risk of a multi-plant explosion, as the two classes of incidents arise from substantially different causes. Lost-time accidents tend to arise from the activities of people in their workplaces, such as the incorrect use of tools. Major accidents tend to depend on large inherent hazards, such as the potential energy of large elevated structures, the combustion energy of fuels and chemicals, the toxicity of some materials, or the kinetic energy of large rotating machines.

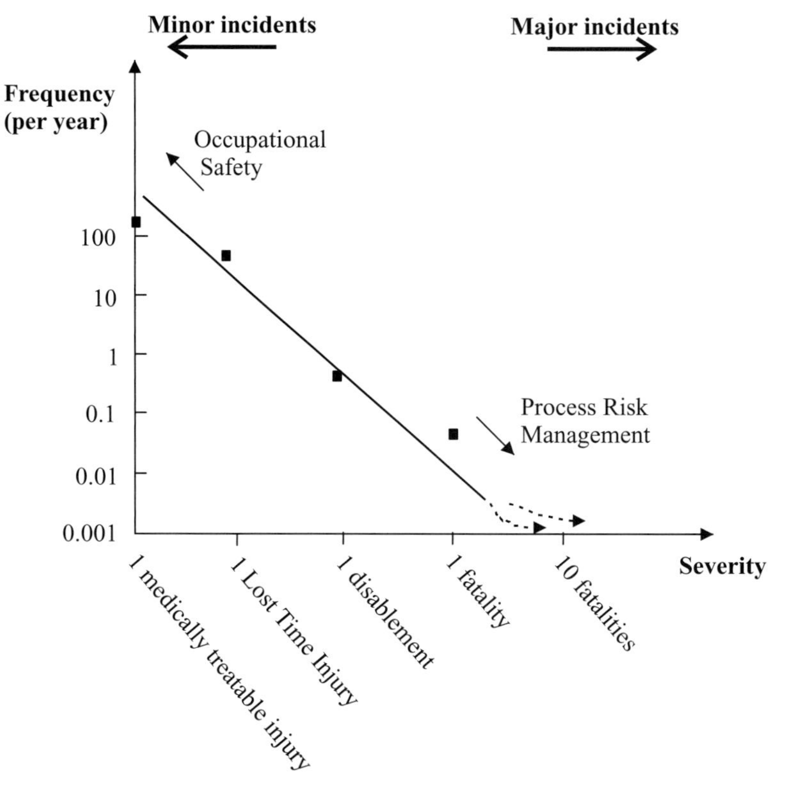

Figure 2.1 Frequency of accidental incidents for a (hypothetical) average organization (Source: based on Tweeddale, 2003).

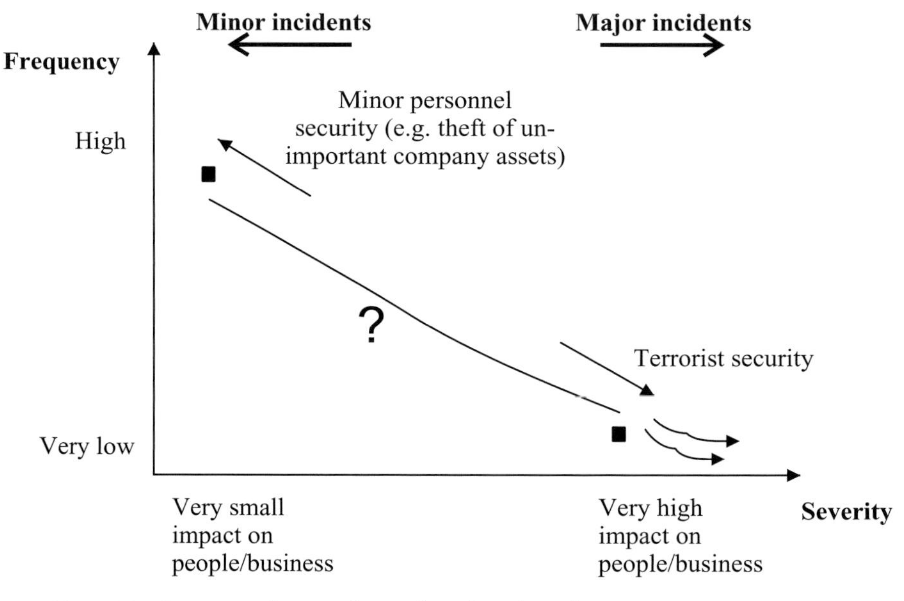

Figure 2.2 Frequency of security losses for a (hypothetical) average organization.

As depicted in Figure 2.1, the frequency of catastrophes on-site of a chemical process plant is very low, which leads to a very slow awareness of both politicians and captains of industry to pro-actively approaching all accidents (including low-frequency accidents) by developing adequate safety regulations and safety-management systems. It was only in the 1990s – after a series of major single plant accidents that took place – that the concept of a "plant safety-management system" was truly accepted by all stakeholders of the chemicals sector and became common practice in the chemical industry.

However, the cross-company disaster frequency is (fortunately) even lower, and as a result, to date, no multi-plant safety-management system or multi-plant-safety culture jointly implemented by a group of chemical plants exists to prevent multi-site accidents within a chemical cluster.

2.4
The Security-Risk Spectrum

Depending on the type of security risk in question, a distinction between (high-frequency, low-consequence) and (low-frequency, high-consequence) security losses can be made (see Figure 2.2).

In the pre-9/11 era, management of industrial security focused on security measures aimed at preventing company personnel from filching company assets, preventing theft (internal as well as external theft, for example, by company employees or by drug dealers), or at preventing information losses and attacks on company software. Without a doubt, the 11 September 2001 attacks formed the principal reason behind the corporate establishment's desire to focus attention on terrorism. As Alexander (2004) explains, in the United States during the period 1996–2000, the annual Pinkerton survey of leading corporations never placed terrorism in its top-ten list of security threats. In 2002, terrorism was ranked third. Similar figures are likely to be valid for European businesses as well. Chemical companies thus became aware that, should they not spend adequate funds on their security, the risk of becoming a target of terrorism was real.

It is not possible to quantitatively assess the level of "security efficiency" against terror, simply because empirical data is not available and cannot be simulated. Analogous to major safety-related accidents, the frequency of terrorist attacks on-site of a chemical process plant is extremely low, which again leads to a very slow awareness of both politicians and captains of industry to pro-actively approaching such threats by developing an operator security plan (OSP).

Multi-site terrorist attacks are even rarer, and as a result, to date, no multi-plant operator security plan or multi-plant security culture is jointly implemented by a group of chemical plants to prevent multi-site attacks within a chemical cluster.

2.5 Multi-Plant Chemical Risks

2.5.1 Domino Effects

A wide variety of chemical safety risks, respectively chemical security risks, exists. The most dangerous ones (situated in the very high severity range) are called *domino risks,* a term by which the potential for a knock-on interaction between groups of installations in the event of an accident at one of the installations is connoted. This mechanism is referred to as "domino", "escalation", "interaction" or "knock-on". Domino risks can either be safety risks, implying having an accidental nature, or security risks, implying having an intentional nature. It is obvious that such events are either way very complex phenomena. Domino risks or the risks associated with domino effects (i.e. domino accidents), have a very high destruction potential. The notion is more accurately defined in Section 2.5.2. The study of domino effects is performed by investigating the different successive accidents, so-called *domino events,* which constitute a domino effect (Delvosalle, 1996; Lees, 1996; Council Directive 96/82/EC, 1997).

While they have been recognized for a long time, the literature remains scarce and vague about the domino effect subject. There is no generally accepted definition of what constitutes domino effects, although various authors have provided suggestions. Table 2.1 presents an overview of current definitions identified in a review of relevant documents.

The generalized definition provided by Delvosalle (1998) is a very generic definition and includes all the different categories of domino effects. According to this definition, a domino effect implies a primary accident concerning a primary installation (this event might not be a major accident), inducing one (or more) secondary accident(s), concerning secondary installation(s). This (these) secondary accident(s) must be (a) major one(s) and must extend the damages of the primary accident.

It should be noted, however, that some definitions from Table 2.1 cannot be applied to all conceivable escalation events, but that these definitions are restricted to a specific type of domino effects, that is domino effects involving more than one plant. Different types of domino effects can thus be distinguished. Whereas "internal domino effects" denote an escalation accident happening inside the boundaries of one chemical plant, "external domino effects" indicate one or more knock-on events happening outside the boundaries of the plant where the domino effect originates, as a direct or as an indirect result (see also Section 2.5.2).

Although the consequences of external domino effects can be devastating, this phenomenon has so far attracted very little attention of prevention managers in existing chemical clusters. The reason for this rather strange observation is twofold. On the one hand, modeling of domino effects is highly complex. On the other hand, the probability of domino accident events is extremely low. In order to assess domino-effect consequences, deterministic models have to be used in

Table 2.1 Non-exhaustive list of domino-effect definitions.

Author(s)	Domino effect definition
HSE (1984)	The effects of major accidents on other plants on the site or nearby sites.
Bagster and Pitblado (1991)	A loss of containment of a plant item that results from a major incident on a nearby plant unit.
Lees (1996)	An event at one unit that causes a further event at another unit.
Khan and Abbasi (1998b)	A chain of accidents or situations when a fire/explosion/missile/toxic load generated by an accident in one unit in an industry causes secondary and higher order accidents in other units.
Delvosalle (1998)	A cascade of accidents (domino events) in which the consequences of a previous accident are increased by the following one(s), spatially as well as temporally, leading to a major accident.
CCPS (2000b)	An accident that starts in one item and may affect nearby items by thermal, blast or fragment impact.
Vallee *et al.* (2002)	An accidental phenomenon affecting one or more installations in an establishment that can cause an accidental phenomenon in an adjacent establishment, leading to a general increase in consequences.
Council Directive 2003/105/EC (2003)	A loss of containment in a Seveso installation that is the result (directly and indirectly) from a loss of containment at a nearby Seveso installation. The two events should happen simultaneously or in very fast subsequent order, and the domino hazards should be larger than those of the initial event.
Post *et al.* (2003)	A major accident in a so-called "exposed company" as a result of a major accident in a so-called "causing company". A domino effect is a subsequent event happening as a consequence of a domino accident.
Cozzani *et al.* (2006)	Accidental sequences having at least three common features: (i) a primary accidental scenario, which initiates the domino accidental sequence; (ii) the propagation of the primary event, due to "an escalation factor" generated by the physical effects of the primary scenario, that results in the damage of at least one secondary equipment item; and (iii) one or more secondary events (i.e. fire, explosion and toxic dispersion), involving the damaged equipment items (the number of secondary events is usually the same of the damaged plant items).
Bozzolan and Messias de Oliveira Neto (2007)	An accident in which a primary event occurring in primary equipment propagates to nearby equipment, triggering one or more secondary events with severe consequences for industrial plants.
Gorrens *et al.* (2009)	A major accident in a so-called secondary installation that is caused by failure of a so-called external hazards source.
Antonioni *et al.* (2009)	The propagation of a primary accidental event to nearby units, causing their damage and further "secondary" accidental events resulting in an overall scenario more severe than the primary event that triggered the escalation.

combination with probabilistic models. The main problem about deterministic modeling arises from the transient nature of the events. The difficulties in the application of probabilistic models are in particular due to the fact that the original input data for the probabilistic analyses are often missing.

In this book, external domino effects are of particular interest because such accidents involve different companies situated within a chemical cluster. It is, however, obvious that such multi-plant domino effects make for one of the most complex accident types existing to date to study and to tackle within an industrial setting.

As already mentioned, (internal and external) domino effects either happen by chance, or they can be deliberately induced. Despite the fact that all domino accidents that have happened thus far are believed to have been accidental, companies forming a chemical cluster such as the very large chemical clusters of Houston (USA), Antwerp (Belgium) and Rotterdam (The Netherlands), should be aware of the possibility of someone intentionally causing a domino effect.

2.5.2
Domino-Events Categorization

A classification of domino events into the various types that may occur is proposed. In this way, it is possible to unambiguously identify the character of the domino event under consideration. The various domino-event types are explained in Table 2.2 using four different parameters.

From the definitions of a direct or indirect domino event, it is not possible to deduce how many domino events have happened before the event under consideration. For this purpose, the concept of *domino cardinality* is introduced. This is a term used to indicate the domino event link number in a sequence of domino events, starting from the initiating event, with domino cardinality "0".

These categorizing definitions may be illustrated by considering a hypothetical domino accident providing a good overview of the different aspects of domino events. The example is summarized in Figure 2.3.

Consider installation *B.1* of company *B* as the lift-off of the domino effect (e.g., assume a gas leak) having cardinality zero (being the origin of the domino effect). As a result, this minor incident develops hypothetically into an escalation accident in the following way. The gas leak gives rise to a gas cloud inside installation *B.1*. At a certain moment, the leaking gas cloud is ignited, resulting in a temporal UVCE[1] at the installation itself (*De1*) and, due to heat from the explosion, in a major fire at a nearby installation (*De6*). The radiation caused by the fire resulting from the UVCE results 15 minutes later in another explosion (*De2*, spatial) affecting an installation unit (*A.1*) situated on the premises of a nearby plant *A*. The major fire in unit *B.2* leads to fire and heavy overheating in the installation items *B.3* and *B.4*, which are both severely damaged at approximately the same time (*De7* and *De8*). A heat-resulting flashfire (*De9*) in installation *B.4* destroys half the

1) A UVCE is a major accident scenario (i.e. unconfined vapor-cloud explosion).

Table 2.2 Categorization of domino events.

Categorization of domino events		
Type number	**Instances of type**	**Definition of type**
Type 1	Internal	Begin and end of the escalation vector characterizing the domino event are situated inside the boundaries of the same chemical plant.
	External	Begin and end of the escalation vector characterizing the domino event are not situated inside the boundaries of the same chemical plant.
Type 2	Direct	The domino event happens as a direct consequence of the previous domino event.
	Indirect	The domino event happens as an indirect consequence of a preceding domino event, not being the previous one.
Type 3	Temporal	The domino event happens within the same area as the preceding event, but with a delay.
	Spatial	The domino event happens outside the area where the preceding event took place.
Type 4	Serial	The domino event happens as a consequent link of the only accident chain caused by the preceding event.
	Parallel	The domino event happens as one of several simultaneous consequent links of accident chains caused by the preceding event.

installation within the hour. The *De2*-explosion causes several fires in installation *A.1*, in turn leading to a classic BLEVE mode (*De3*). Due to *De3*, another installation (*A.2*) of company *A* gives rise to an explosion and the subsequent missiles cause minor damage to installations *A.7* (*De4*) and *B.5* (*De5*).

The different exemplary domino events can then be categorized as in Table 2.3.

Another way of representing this hypothetical domino effect is by way of a domino-event tree. Figure 2.4 illustrates such a tree.

Domino events characterized with cardinality 1 are also called "primary domino events", whereas cardinality 2 refers to secondary domino events, cardinality 3 to tertiary domino events, *etc.* An external domino effect indicates a multi-plant domino effect and represents an escalating accident involving more than one chemical plant (note that "multi-plant" thus simply indicates the involvement of more than one plant). Hence, according to the categorization of domino events (Table 2.2), a multi-plant accident can be considered to be a domino effect typified by external and spatial characteristics.

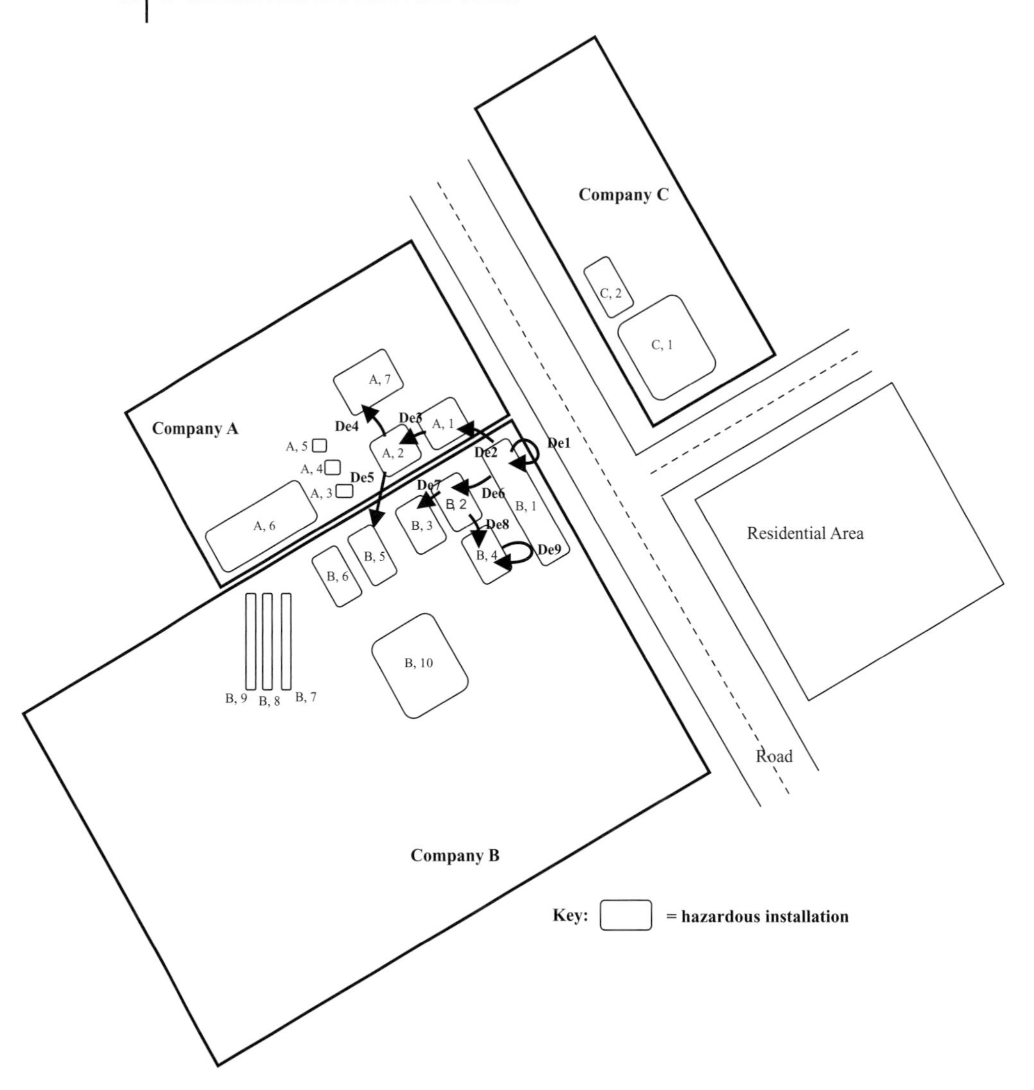

Figure 2.3 Hypothetical domino effect illustrating the constituting domino events' ("De") complexity.

2.5.3 Domino Effects in the Past

Available reports on domino effects, for example, Pietersen (1986), Mahoney (1990), Pietersen (1990), Fievez (1996), Lees (1996), Kletz (1998), and Khan and Abbasi (2001), describe, often qualitatively, past single-plant and multi-plant chemical knock-on accidents. The worst such accident – in terms of death toll – occurred in Mexico City on November 19, 1984, and claimed 650 lives and resulted in more than 6400 injuries. Table 2.4 presents an illustrative list of some of the major domino effects that have occurred since 1917.

Table 2.3 Type of hypothetical exemplary domino events depicted in Figure 2.3.

Type		De0	De1	De2	De3	De4	De5	De6	De7	De8	De9
1	Internal		■		■	■		■	■	■	■
	External			■			■				
2	Cardinality	0	1	2	3	4	4	1	2	2	3
3	Temporal		■								■
	Spatial			■	■	■	■	■	■	■	
4	Serial			■	■						■
	Parallel		P1			P3	P3	P1	P2	P2	

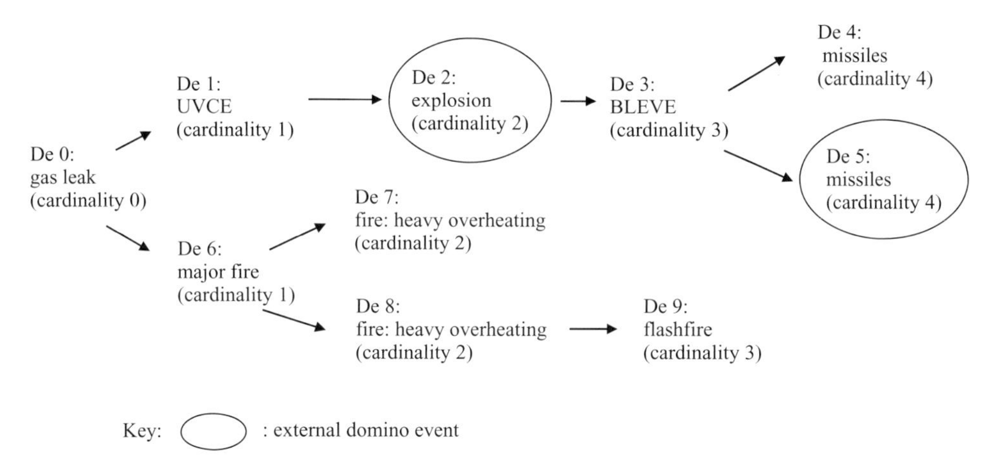

Figure 2.4 Domino event tree of the hypothetical domino effect of Figure 2.3.

The Buncefield disaster, a multi-plant domino effect that happened near London (United Kingdom) in 2005, is said to be the largest fire accident of peacetime Europe. The dense pall of smoke rose as high as 3000 m over the burning storage tanks where it originated from; the plume was so vast it appeared in satellite images of the scene. This accident has led to estimated financial losses of approximately 2 billion Euros, which represent the largest financial losses of any domino effect that has taken place thus far.

Most of the accidents mentioned in Table 2.4 are internal domino effects. However, accidents involving several plants (such as the Hemel Hempstead accident or the Mexico City accident) do occur, and their human and economic loss potential is often many times greater than that of single-plant escalation accidents. This observation certainly justifies the existence and the importance of scientific and professional studies helping to prevent devastating multi-plant accidents, besides all studies carried out, and all requirements and measures in place in chemical plants to control and to manage internal domino effects.

Table 2.4 Non-exhaustive list of accidents involving domino events.

Location	Chemical involved	Date	Number of fatalities	Number of injuries
Ashton, UK	Explosives	1917	46	>120
Nova Scotia, Canada	Explosives	1917	1800	8000
Neukirchen, Germany	Town gas	1933	65	>100
Saint Chamas, France	Explosives	1936	53	150
Ludwigshafen, Germany	–	1943	57	439
Cleveland, USA	LNG	1944	128	200–400
Texas City, USA	Ammonium nitrate	1947	552	>3000
Ludwigshafen, Germany	Dimethyl ether	1948	207	3818
Port Newark, USA	LPG	1951	0	14
Bound Brook, USA	Phenolic resin	1952	5	21
Lake Port, USA	LPG	1954	4	–
Whiting, USA	Naphta	1955	2	40
Montreal, Canada	Butane	1957	1	–
Meraux, USA	Diesel oil	1957	1	–
Boron, USA	LPG	1958	1	–
Signal Hill, USA	Oil	1958	2	18
Deer Lake, USA	LPG	1959	11	10
Mac Kittrict, USA	LPG	LPG	–	2
Pampa, USA	Isopentane	1962	–	1
Doe Run, USA	Ethylene oxide	1962	1	21
Marietta, USA	Benzene	1962	1	3
Attleboro, USA	Vinyl chloride	1964	7	40
Nigata, Japan	Natural gas	1964	3	–
Louisville, USA	Mono vinyl acetylene	1965	12	8
Feyzin, France	Propane	1966	18	81
Lake Charles, USA	Isobutene	1967	7	13
Port Arthur, USA	Petrol	1968	3	5
Pernis, The Netherlands	Light hydrocarbons	1968	2	85
Leverkusen, Germany	Urea	1968	1	14
Dormagen, Germany	Ethylene oxide	1969	1	40
Laurel, USA	LPG	1969	2	976
Glendora, USA	Vinyl chloride	1969	–	1
Repcélak, Hongary	Purified CO2	1969	9	15
Round Brook, USA	Nitrophenol	1969	1	1
Long Beach, USA	Petrol	1969	1	83
Texas City, USA	Butadiene	1969	13	5
Crescent City, USA	LPG	1970	0	66
Ludwigshaven, Germany	Propylene	1970	5	3
Baton Rouge, USA	Ethylene	1971	–	21
Houston, USA	Vinyl chloride monomer	1971	1	5
Longview, USA	Ethylene	1971	4	60
Rio de Janeiro, Brazil	LPG	1972	37	53
Duque de Caxias, Brazil	LPG	1972	39	51
Kingman, USA	LPG	1973	13	95
Beaumont, Texas, USA	Hydrocarbons	1974	2	10
Beek, The Netherlands	Propylene	1975	14	107

Table 2.4 *Continued*

Location	Chemical involved	Date	Number of fatalities	Number of injuries
Guayaquil, Ecuador	LNG	1976	–	>50
Puebla, Mexico	Vinyl chloride monomer	1977	1	3
Texas City, USA	LPG	1978	7	10
Tacoma, USA	Hydrogen	1979	–	3
Ras Tanura, Saudi Arabia	Petrol	1979	2	6
Prioli, Italy	Cumene	1979	1	–
Milligan, USA	Ammonia, acetone, *etc.*	1979	0	14
Deer Park, USA	Vacuum distillate	1980	3	12
Borger, USA	Light hydrocarbons	1980	0	41
Montana, USA	Chlorine	1981	17	1000
Stailybridge, UK	Hexane	1981	1	1
Caracas, Venezuela	Fuel oil	1982	175	500
Milford Haven, UK	Crude oil	1983	–	20
Houston, USA	Methyl bromide	1983	2	>1
Mexico City, Mexico	LPG	1984	650	6400
Romeoville, USA	Propane, butane	1984	17	31
San Antonio, USA	ammonia	1985	4	23
Naples, Italy	Oil	1985	4	170
Petal, USA	LPG	1986	–	12
Lyon, France	Xylene and polyisobutylene	1987	2	14
Antwerp, Belgium	Ethylene oxide	1987	–	14
Port Herriot, France	Oil	1987	2	8
Genoa, Italy	Methanol and hexane	1987	4	1
Bombay, India	Benzene	1988	35	16
Brisbane, Australia	Detergent	1988	–	30
Jonova, Latvia	Liquefied ammonia	1989	7	57
Antwerp, Belgium	Ethylene oxide	1989	2	5
Pasadena, USA	Isobutene	1989	23	124
Al Hillah, Iraq	Explosives	1989	19	–
Nagothane, India	Ethane and propane	1990	31	63
Stanlow, UK	Different chemicals	1990	1	5
Bradford, UK	Azodiisobutyronitrile	1992	–	33
Shenzhen, China	Ammonium nitrate	1993	15	141
Milford Haven, USA	Hydrocarbons	1994	–	26
Sioux City, USA	Nitric acid	1994	5	18
New Delhi, India	Nitrocellulose	1994	8	2
Ueda, Japan	Gasoline	1994	1	3
Kucove, Albania	Crude oil	1995	1	4
Martinez, USA	Hydrogen	1996	–	2
Burnside, USA	LPG	1997	2	2
Vishakhapatnam, India	LPG	1997	60	–
Zamboanga, Philippines	Hydrocarbons	1997	1	6
Albert City, USA	Propane	1998	2	7
Longford Victoria, Australia	Hydrocarbons	1998	2	8
Texas, USA	Propane	2000	2	1
Ohio, USA	Different chemicals	2001	–	17
Pennsylvania, USA	Explosives	2001	1	3

Table 2.4 *Continued*

Location	Chemical involved	Date	Number of fatalities	Number of injuries
Roncador, Brazil	Hydrocarbons	2001	10	–
Delaware, USA	Sulfuric acid	2001	1	8
Toulouse, France	Ammonium nitrate	2001	30	>5000
Mexico City, Mexico	Alcohol and methanol	2001	–	17
Mississippi, USA	Powdered rubber	2002	4	8
Kuwait	Crude oil	2002	4	–
Skikda, Algeria	LNG	2004	27	74
Chongqing, China	Chlorine	2004	9	3
Vadodra, India	Slurry	2004	2	16
Zahedran, Iran	Gasoline	2004	90	114
Neyshabur, Iran	Sulfur, gasoline, fertilizers, cotton wool	2004	328	460
Texas City, USA	Hydrocarbons	2005	15	170
Buncefield, UK	Various fuels	2005	0	43
Fort Worth, USA	Different chemicals	2005	0	4
Arak, Iran	2-ethyle hexanol	2008	30	50
Yizhou City, China	Polyvinyl acetate and other chemicals	2008	16	57
Nagothane, India	Polymers	2008	4	15
Yerevan, Armenia	Synthetic rubber	2009	3	24
Lalbagh, Bangladesh	Polymers	2009	6	6
Torkham, Pakistan	Crude oil	2009	–	60
Badami Bagh, Pakistan	Different chemicals	2009	>12	–

Source: based on Mahoney (1990); Fievez (1996); Khan and Abassi (1998a); ES1 (MARS, 2009); Abdolhamidzadeh (2009).

2.5.4 Multi-Plant Chemical-Risk Measurement

A criterion that relates to the chemical risk to any particular individual is insufficient on its own. There is an important difference between an incident that might occur with a probability of one in a million per year fatally injuring one person and an incident with the same probability that could fatally injure 1 000 people. For this reason, the "societal risk" or "group risk" criterion is introduced. The best available technology for studying this Societal Risk is full scope application of quantitative risk assessment (QRA). Such studies lead to a relationship between the number of fatalities (N_F) that can follow a major accident, ranging from one to some maximum value (N_{max}) and the frequency (f) at which that number of fatalities is estimated to occur. This relationship or more usually the corresponding relationship involving F, the cumulative frequency of events having N_F or more fatalities, is usually presented as a graph as depicted in Figure 2.5.

Societal risk is designed to display how risks vary with changing levels of severity. For example, a hazard may have an acceptable level of risk for just one fatality,

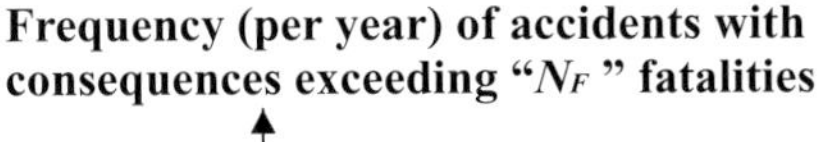

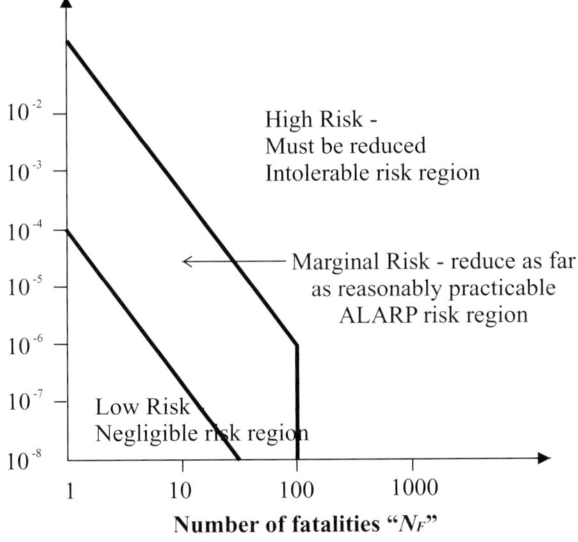

Figure 2.5 Example of approach to defining societal-risk criteria.

but may be at an unacceptable level for 100 fatalities. F–N_F curves may also be used to show the consequences of real accidents that have occurred at installations using dangerous substances. Such plots have the capacity to indicate on the basis of past event observations the individual or social dimension of possible future accidents. One measure of the societal risk from an installation could be obtained by calculating the fatal accident rate (FAR) of the number of fatalities per year from accidents involving dangerous substances. If the F–N_F curve of the installation is known then the value of FAR can be calculated as follows:

$$\mathrm{FAR} = \sum_{N_F=1}^{N_{max}} f(N_F) \cdot N_F$$

However, using the FAR as a criterion for societal risk can be criticized for it does not include an allowance for aversion to multi-fatality accidents. It gives equal weight to the frequencies and consequences of accidents. By not distinguishing between one accident causing 50 fatalities and 50 accidents each causing one fatality over the same period of time, the FAR fails to reflect the importance society attaches to major accidents (Ale, 2005). Moreover, the difficulty in defining societal risk criteria for individual industrial installations within different plants of a chemical cluster still remains. Several installations from nearby companies in the same industrial cluster may each generate a low level of societal risk, whereas their combined societal risk might fall within the high-risk zone of the chart of Figure 2.5, if these installations were all to be grouped for the purposes of the calculation.

In chemical enterprises encountering numerous hazards with severe potential consequences, computer programs are used to calculate the safety risk levels on a topological grid, after which they are used to plot contours of risk on the grid. These contours are used to display the frequency of exceeding excessive levels of hazardous exposure. For example, Reniers *et al.* (2006) indicate available software tools that will prepare contours, for example, for the frequency of exposure to nominated levels of heat radiation, explosion overpressure and toxic-gas concentration. Thus, these software tools provide insight into the possible scale of a disaster and the possibility of its occurrence, but they do not offer adequate information – be it safety related or security related – for the *optimal prevention* of catastrophic accidents.

If the number of events observed in the past is not sufficient to estimate significant frequency values, as in the case of domino effects, a simple histogram plotting the absolute number of past events *versus* a certain type of consequence is often used instead of a risk curve. However, multi-plant accidental or intentional external domino effect predictions are extremely difficult to make in chemical clusters simply because of the lack of sufficient data. Although it is thus not possible to take highly specific multi-plant precaution measures based on statistic predictive information, multi-plant chemical-risk management is actually essential to prevent large-scale multi-plant chemical accidents and therefore should in one way or another be fully incorporated in chemical clusters worldwide.

2.6 Multi-Plant Chemical-Risk Management

In spite of the destructive capability of external domino effects, and the wide array of potential multi-plant chemical risks that many chemical clusters face worldwide, such phenomena have received very limited attention compared with other aspects of risk assessment. Nonetheless, despite all efforts that have been undertaken by the chemical industries, especially regarding the development of safety-risk-assessment techniques and safety-risk-management programs, major chemical accidents still happen. In fact, Khan and Abbasi (1999) analyzed a number of major accidents with a view to understanding the damage potential of various types of disasters. The study concludes that although the number of accidents (in general) has declined, the extent of damage per accident has increased substantially. Thus, with an increase in the density of chemical plants in a cluster and rising population densities worldwide, the probabilities of multi-plant chemical accidents increase sharply. Therefore, there is an urgent need for installing or increasing multi-plant safety-risk management, especially in circumstances where large chemical clusters and densely populated areas coexist. Considering the present importance of terrorist threats worldwide, these observations can easily be extended to security-risk management requirements for preventing intentional attacks possibly leading to multi-plant chemical accidents.

An important aspect of solid risk-management strategy is the carrying out of (safety and/or security) risk-assessment studies. Because of the complex nature of domino-effect risks, it is very difficult, even for experts, to assess such events. Therefore, a variety of computer automated tools have been developed to determine the possibility of accidental (safety-related) domino effects. Based on these highly complex domino-safety-risk-assessment studies (including multi-plant chemical risk assessments), managerial decisions are taken as efficiently and effectively as possible, taking into account both safety and economic constraints. It should, however, be noted that chemical companies are often not obliged to carry out such domino risk assessments, and in the case all neighboring companies would indeed carry out these complex studies, different software packages are often used, characterized by different inputs and outputs, making it very difficult to jointly carry out multi-plant chemical risk studies in current industrial settings. Furthermore, to the best of the author's knowledge, no software is available to determine the vulnerability of an industrial area for deliberately intended attacks leading to multi-plant chemical accidents.

Ongoing research is thus necessary to control multi-plant chemical hazards and to prevent the incidence of (accidental as well as deliberate) major accidents becoming greater. It is crucial to investigate how to establish and to elaborate cross-company chemical-risk management for external domino effects that are related to both safety risks and security risks.

Although a multi-plant chemical-risk-management strategy would substantially increase safety and security in chemical clusters, no such strategy exists. Moreover, no clear economic incentives exist to urge companies within chemical clusters to jointly develop such a multi-plant risk-management strategy. To this end, the next section introduces the concept of "hypothetical benefits" and associates the term with multi-plant chemical risks.

2.7
Hypothetical Benefits Associated with Multi-Plant Chemical Risks

Investing in multi-plant prevention measures is in many companies considered a cost that contributes little to corporate results or financial performance. As a consequence, the vast majority of companies fail to appreciate the benefits offered by investing in multi-plant safety and security. One of the main reasons for this lack of interest in multi-plant safety and security benefits from corporate financial analysts is that reducing financial loss as a result of off-site incident avoidance through the implementation of preventive measures is an extremely theoretical concept and quantification is very difficult. A survey conducted by Reniers (Reniers and Soudan, 2003) was (among others) aimed at measuring company efforts to undertake this economic exercise. It transpired that, of 24 chemical companies questioned, not one respondent had taken time to calculate in detail the hypothetical benefits resulting from implemented safety and security measures within the company.

Despite the doubt surrounding the details and accuracy of the quantification results of such benefits, it is still worth specifying the theoretical reduction in loss as a result of multi-plant safety and security investments. By accessing more accurate quantitative figures, senior management gains a better insight into investing in multi-plant safety and security and is able to take more objective safety and security risk decisions where several plants are involved at once.

A multi-plant accident is linked to all manner of high direct and indirect costs. It can thus be stated that by implementing a multi-plant safety and security policy and by taking adequate preventative safety and security measures very important costs can be avoided, namely all costs associated with multi-plant accidents that have never occurred. *Non-occurring* multi-plant accidents (in other words the prevention of such accidents) thus result in avoidance of a number of very high costs and thus create very high "hypothetical benefits" (Reniers and Audenaert 2009d). Hence, the costs of preventative measures must be weighed against hypothetical benefits resulting from the *non-occurrence* of an accident or resulting from sound safety and health policies within a cluster of companies.

The optimum degree of safety required to prevent cross-company losses is open to question, both from a financial and economic point of view and from a policy point of view. Developing and executing a sound multi-plant prevention policy involves costs but the avoidance of large-scale accidents and damage leads, on the other hand, to high hypothetical benefits. Consequently, a chemical cluster that deals with multi-plant safety and security must try to establish an optimum between, on the one hand, the investment (costs) and the resulting degree of safety realized as a result of the investment and, on the other hand, the avoidance of damage of any kind (hypothetical benefit). The aim should be to balance investments in multi-plant safety and security with hypothetical benefits in such a way that the latter increase thanks to a reduction in potential damage costs. Thus, the total potential costs decrease notwithstanding the safety and security investments.

It is possible to further expand upon costs and benefits of accidents in general in terms of the "degree of safety and security" (S). The degree of safety and security is the "measure of safety and security" of a company in question. The theoretical S can therefore vary between a large measure of risk ($S = (0 + \varepsilon)\%$) and a large measure of safety/security ($S = (100 - \varepsilon)\%$), wherein ε assumes a (small) value suggesting that "absolute risk" or "absolute safety/security" in a company is, in reality, not possible. The economic break-even safety point ($\tilde{S}$), namely the point at which the cost of safety and security measures (C_S) equals the resulting benefits (B_S), can be represented in graph form (see Figure 2.6). The graph (which was derived on the basis of confidential semi-qualitative company data) distinguishes between two cases: major accidents[2] and minor accidents.[3] Four conditions should be met for an event to be characterized a *major* accident: (i) hazardous materials are involved, (ii) the accident is linked to an unexpected incident during industrial activity,[4] (iii)

2) Major accidents are, for example, multi-fatality accidents, accidents with huge economic losses, accidents with an off-site impact such as multi-plant chemical accidents, *etc.*

3) Minor accidents are for example accidents resulting in the inability to work for several days, accidents requiring first aid, *etc.*

4) An "unexpected incident" is defined as an abnormal, uncontrollable (escalating) development during the activity (or storage) in question.

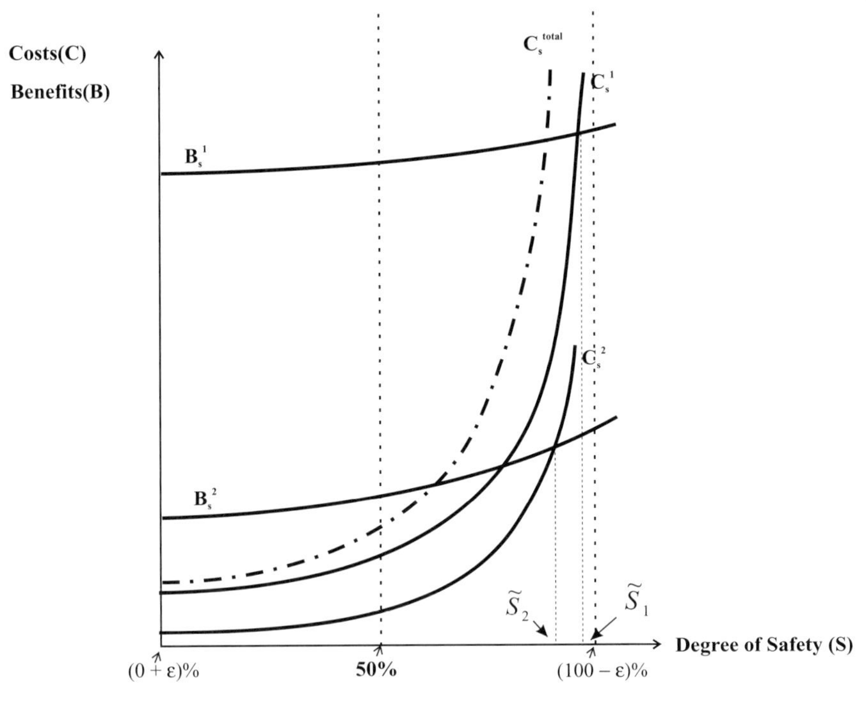

Figure 2.6 Graphic representation of the economic break-even safety points relating to a major accident scenario (with index 1) and to a minor accident scenario (with index 2).

the event results in extreme danger to people and/or the environment and (iv) one of the characteristics of the accident is an explosion, a major fire and/or the emission of a poisonous gas cloud. Any accident in which one of these criteria is not represented is deemed a *minor* accident. This distinction must be emphasized due to the fact that there is a considerable difference in the costs and benefits relating to the two different sorts of accident. In the case of minor accidents the hypothetical benefits resulting from *non-occurring* accidents are considerably lower than in the case of major accidents. This means essentially that the company reaps greater potential financial benefits from investing in the prevention of major accidents than in the prevention of minor accidents. When calculated, costs related to preventive measures for major accidents are, in general, also considerably higher than for minor accidents. Major accidents are thus mostly prevented by means of (expensive) technical studies and the furnishing of (expensive) technical equipment within the framework of process safety. Minor accidents are therefore more associated with the protection of the individual, first aid, the daily management of safety codes, *etc.*

As the company invests more in safety the degree of safety will also increase. Thus, as S increases, a rising C_S curve is obtained. Moreover, the higher the degree of safety the more difficult it becomes to improve upon this (i.e. to increase it) and the curve depicting investments in safety thereafter displays asymptotic characteristics. As more in safety is invested from $S = (0 + \varepsilon)\%$, higher benefits are obtained as a result of *non-occurring* accidents. These curves will display a much more level trajectory due to the fact that marginal investments do not produce large additional benefits. The benefits curve and the costs curve dissect at a break-even safety point $\tilde{S}$. If there are greater investments in safety following this point, benefits will no longer balance costs. A different position for the costs and benefits curves, and thus also for $\tilde{S}$, is obtained for the two different types of accident scenario (major accidents and minor accidents).

Figure 2.6 illustrates the qualitative benefits curves[5] for both types of accident scenario; these are labelled B_S^1 and B_S^2. The index "1" benefits curve represents major accidents and the index "2" curve represents minor accidents. It is clear that benefits from safety measures (i.e. the avoidance of costs thanks to *non-occurring* accidents) B_S^2 and costs of safety measures C_S^2 relating to minor accidents are considerably lower than those relating to major accidents (curves B_S^1 and C_S^1). The cost/benefit break-even safety point for minor accidents ($\tilde{S}_2$) is likewise lower than for major accidents ($\tilde{S}_1$).

Figure 2.6 also shows that in the case of major accidents where a company is subject to extremely high financial damage, (i) the hypothetical benefits resulting from the avoidance of accidents such as these are even higher and (ii) the break-even point must be located near the $(100 - \varepsilon)\%$ degree of safety limit. This supports the argument that in the case of major accidents all necessary safety investments can be justified, notwithstanding the size of the costs. The costs avoided (should such accidents never occur) or the so-called hypothetical benefits are certainly (nearly) always higher than the costs of the safety measures required. Thus, for major accidents it is possible to state that no real cost-benefit comparison is required but that a degree of safety for major accidents of $(100 - \varepsilon)\%$ must be guaranteed in spite of the cost of corresponding safety measures. It goes without saying that safety costs aimed at preventing major accidents also serve partly to prevent minor accidents and *vice versa*. Similarly hypothetical benefits relating to major and minor accidents cannot be separated.

For a single company to be successful in the long-term in occupational health and safety it is very important to ensure that the safety risk posed by a process system is always below generally accepted company levels. This can be achieved by developing a solid plant safety-management system. Such a system aims to adopt a systematic and pro-active approach to evaluation and management of the plant and its products throughout its life, considering safety features throughout process selection, process design, plant realization, commissioning, production and decommissioning. Arrangements are made to ensure that the means provided for safe operation of the industrial facility are properly designed, constructed, tested, operated, inspected and maintained, and that persons working on the site are properly trained. However, the implementation of an individual plant safety-

5) In this case, qualitative benefit curves are general curves that were derived, on the basis of semi-qualitative data, and that *exact curve positions* can vary from situation to situation and from company to company.

management system is not an optimal approach to tackle major multi-plant hazards in a complex industrial surrounding of several chemical plants and is not always sufficient to prevent multi-plant major accidents from happening.

Suppose that a multi-plant chemical accident within company *A* is the result of one or more accidents that took place at an adjacent company *B*. Hence, for such an accident, no direct *prevention* costs can be taken in company *A*. Company *A* can only *protect* itself against possible cross-plant accident scenarios outgoing from company *B*. Hence, to decrease multi-plant chemical risks, according to present risk-management working practices a chemical plant takes all kinds of preventive and protective measures against possible internal-originating major accidents (indirectly decreasing the likelihood and the consequences of an off-site impact) and it takes protective measures against potential external-originating accidents. This cross-company safety and security strategy currently used within chemical companies and clusters worldwide is by no doubt economically sub-optimal compared with looking at a chemical cluster in an integrated (holistic) way and taking *preventive as well as protective* measures against multi-plant chemical risks by neighboring plants *together* (in collaboration).

The reasons for investing in multi-plant security-risk management are essentially the same as those for investing in multi-plant safety-risk management. However, avoidance or minimization of the security risks is dealt with by different (security) risk-management strategies, entailing different costs. Security-risk management may comprise security-risk assessment, development of security policy, collaboration with local, state, and national law-enforcement agencies and with local emergency planning committees, periodic reassessment of the security plan for physical security, including access control, perimeter protection, intrusion detection, security officers and on-going testing, employee security measures, workplace violence prevention and response, information security, computer security, network security, *etc.*

The existing differences between safety-risk assessments and security-risk assessments are explained in Sections 2.8 and 2.9, respectively.

2.8
Safety-Risk Assessment and Safety-Risk Management

Safety-risk assessment connotes a systematic approach to organizing and analyzing scientific knowledge and information for potentially hazardous activities or for substances that might pose safety risks under specified circumstances (National Research Council, 1994). The overall objective of chemical safety-risk assessment is to estimate the level of safety risk associated with adverse safety effects of the broad set of technical chemical process conditions. By doing so, it supports the ability to deal with minor accidents as well as major accidents through safety-risk management. Hence, safety-risk-based decision making consists of safety-risk assessment and safety-risk management. The former is the process by which the results of a safety-risk analysis (i.e. safety-risk estimates) are used to make

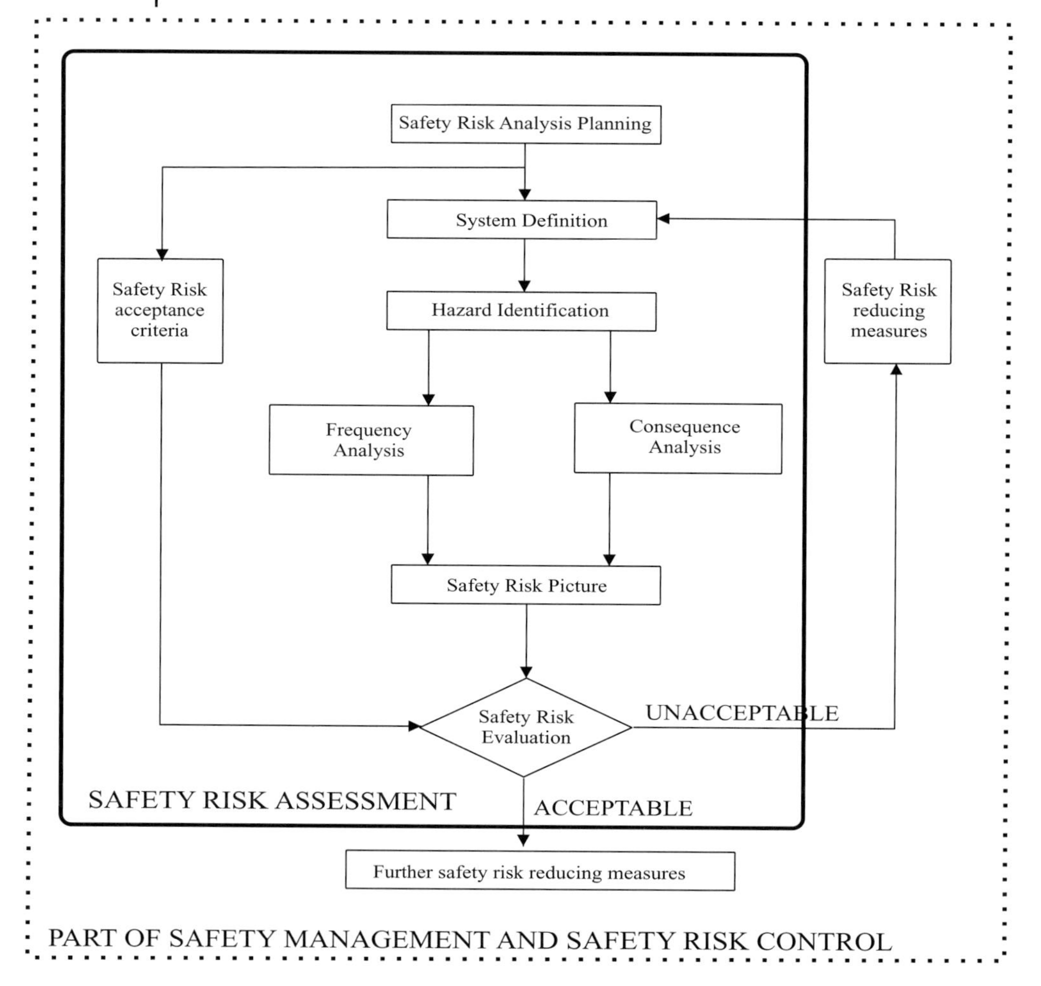

Figure 2.7 Iterative process of safety-risk assessment as part of safety management and safety risk control (Source: based on Kirchsteiger, 2003).

decisions, either through relative ranking of safety-risk-reduction strategies or through comparison with safety-risk targets; whilst the latter consists of the planning, organizing, leading and controlling of an organization's assets and activities in ways that minimize the adverse operational and financial effects of accidental losses upon the organization.

The sequence of the various process steps in assessing the safety risks originating from a specified system, that is, establishing the context, identification, analysis, assessment, management and decision making is very similar at the generic level across different industries and countries. The safety-risk-assessment process is illustrated in Figure 2.7.

In the case of multi-plant safety risks, decision makers are confronted with a variety of concrete approaches and methodologies for implementing the different steps of the generic scheme. Non-uniformity in methods, data and applications thus significantly hamper the widespread use of multi-plant safety risk assessments for decision-making purposes.

In contrast to safety-risk assessment which is rather technical matter, safety-risk management is largely based on company policy and can be seen as a response to perceptions. Therefore, safety-risk management significantly differs across chemical plants mainly as a result of the different values and attitudes towards specific safety risks in different plant-culture contexts.

2.9 Security-Risk Assessment and Security-Risk Management

Security-vulnerability analysis (SVA) connotes a systematic approach to organizing and analyzing scientific knowledge and information concerning the assets that need to be protected in chemical plants, the threats that may be posed against those assets, and the likelihood and consequences of attacks against those assets. Assets can be defined as people, information, and property. Hence, SVA serves to improve a chemical company's understanding of the threats and possible responses that may exist within the plant. As such, SVA forms the basis for establishing a cost-effective security-risk management program suitable to reduce the potential adverse effects of intentionally induced losses upon the company.

A security-vulnerability analysis at the generic level across different industries and countries is presented in Figure 2.8.

Assessing security risks by applying the loop as described in Figure 2.8 requires the creation of a security-management program. Such a program also deals with cost-effectiveness, and promoting security awareness among others. However, managing security risks not only depends on company policy and company perceptions and attitudes towards security risks, but also on neighboring companies' perceptions, as well as on public perception and law enforcement. Therefore, it is crucial to take multi-plant accidents into account when establishing a security-management program. Managing security in an integrated multi-plant context would certainly enhance and optimize the prevention of possible multi-plant intended accidents. It would also lead to an increased public image of the chemical industry's security efforts, and at the same time possibly discourage potential terrorists. If a chemical enterprise or indeed a chemical cluster were to be confronted by a large-scale terrorist attack, the whole industry would be likely to be subject to more stringent security regulations.

2.10 Summary and Conclusions

In this chapter, current insights into chemical safety and security risks and chemical safety and security-risk assessment have been presented. Since chemical

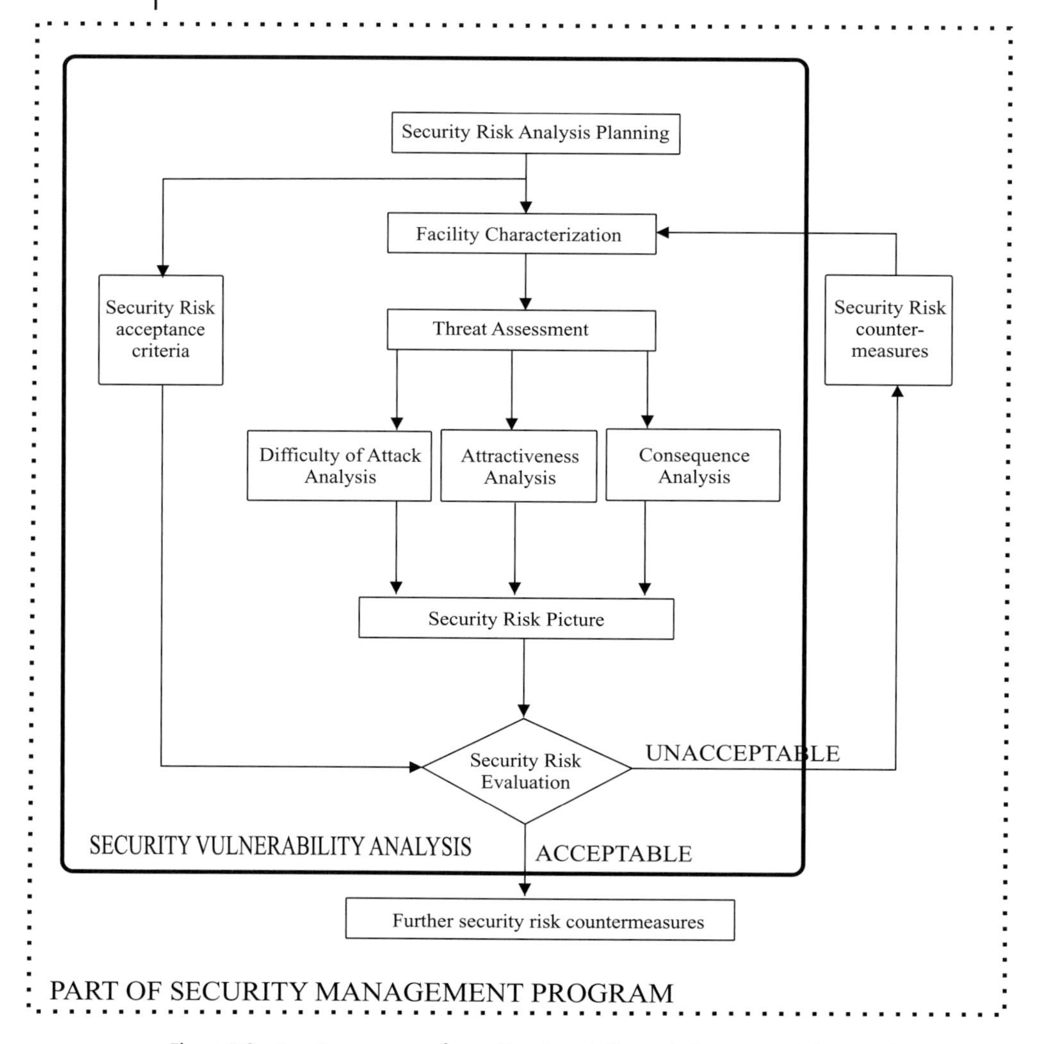

Figure 2.8 Iterative process of security-vulnerability analysis as part of the security-management program (Reniers, 2009b).

industries are evolving towards clusters handling increasing amounts of dangerous substances, the emphasis is on multi-plant chemical risks and on domino effects that can eventually lead to catastrophic multi-plant accidents. In chemical clusters where the concentration of chemical industrial activity is very high within areas of high population density, an integrated tool for efficiently managing multi-plant risks leading to a reliable course of actions to be undertaken, represents a crucial need.

Prior to September 11th 2001, a successful intentional attack (e.g., by terrorists) on a chemical facility was believed to be extremely unlikely (although terrorist threats had already been recognized for a long time). Chemical plant cultures did not take into account security vulnerability analyses, except in very special circumstances. The ramifications of the post-9/11 era include heightened security risk with regard to physical and economic damage. In the chemical industry in particular that is subject to risks possibly leading to major accidents (such as devastating domino effects), the security implications for everyday operations might prove to be very significant in the prevention of incidents intentionally designed to damage.

To date, little or no attention has been paid to the goals and values jointly held by plant management of different chemical companies belonging to the same cluster. These goals and values determine whether or not a healthy cluster organizational culture develops in relation to multi-plant accident prevention (with regard to safety as well as security). This requires a perspective on multi-plant prevention management involving a strategic approach that integrates multi-plant risk assessment, risk prevention and mitigation, risk communication, *etc.*, into a multi-plant's safety-management system on the one hand and into a multi-plant's security-management program on the other hand.

3
A Multi-Plant Safety and Security Culture: The Requirements

3.1
Introduction

The challenges for effective safety and security management and practices differ considerably between individual chemical companies. The risk-management context is a combination of legislation, insurance, certification, company culture and inter-organizational networks. Hence, the safety and security risk research community is – even per plant – broken down into specialized disciplines such as engineering, psychology, management, organizational sociology, economics and law. Although at first there may be little knowledge transfer and collective learning between the members of this multi-disciplinary group, such an approach can be very useful at the plant level if the various members understand and appreciate each other. However, at the multi-company level the differences between all these disciplines are even extrapolated and individual insights are partially steered by company policies and cultures, both factors hampering cooperation enhancement. The result is sub-optimization of resources for improving safety and security in a chemical cluster and a dispersed distribution of societal risk between individual chemical plants.

To prevent (accidental and intentional) cross-company accidents, each individual plant has at best its own decision method for taking precaution measures. However, a predictive model of external domino-accident causation cannot be developed by simply aggregating models that have been developed separately by different individual plants. A cross-company approach driven by a united problem definition and understanding of the primary domino-hazardous processes, must be applied to pro-actively manage external domino risks in a chemical cluster. This can only be realized if companies belonging to a multi-plant area are jointly prepared to recognize the real potential danger of such accidents. Although these accidents may be characterized by an extremely low frequency rate if considered accidental, the probability of a domino event happening as a result of malicious intent might prove considerable. Therefore, in order to acknowledge the devastating consequences of external escalating incidents, foresight or the willingness to anticipate such events before they happen has to be embedded in the conceptualization of multi-plant organizational systems and processes.

Multi-Plant Safety and Security Management in the Chemical and Process Industries. G.L.L. Reniers

ISBN: 978-3-527-32551-1

As pointed out by Tsoukas and Shepherd (2004), a key player has foresight when he has the natural tendency to act in a manner that coherently connects past, present and future. At an aggregated (multi-plant) safety and security level, this happens when plants all individually have a memory in which past incidents are recorded; this memory is shared between the plants participating to some multi-plant safety and security initiative and certain relations between the items stored in the aggregated memory (e.g., incident characteristics, *etc.*) are deciphered that enable the multi-plant initiative to anticipate, to a certain extent, future incidents. Domino-accident prevention employs another, more complex, way of relating past, present and future. Since knock-on incidents are very rare, the cluster has to hypothesize which of these events will take place in the future and work backwards to the present state to decide what would be needed should these predictions come true? It can be suggested that organizing multi-plant safety and security requires the development of a process for institutionalizing safety and security representations, safety and security routines and safety and security sequences of predictable acts and events in a multi-plant context.

However, safety and security continue, in many cases, to be more arts than sciences. The quality of safety and security is largely determined by the skills and talents of individual analysts, and to a lesser extent by the systematic application of accepted tools and techniques. A key factor in explaining this observation is the lack of commonality of safety and security terms, safety and security tools and safety and security techniques. To advance multi-plant safety and security in a systematic manner and to enhance the cross-plant understanding of safety and security terminology, a multi-plant safety and security culture should be established. Two questions that need answering, are (i) how to efficiently encourage companies (and companies' top management) to jointly install such a safety and security culture, and (ii) how such a safety and security culture can in fact be elaborated in industrial practice? The former question is discussed in Section 3.2, whereas the latter question is treated in the sections thereafter.

3.2 Encouraging Companies to Install a Multi-Plant-Safety and -Security Culture

As noted by Hopkins and Hale (2002), a revolution in the general approach to the regulation of safety has taken place. Rather than specifying standards and procedures to enhance safety, legislation should specify the goal itself, safety, and require employers to ensure the safety of their workers. In principle, it should be up to the employer to decide how best to achieve this goal. Although the goal is safety, employers cannot be expected to guarantee the absolute safety of their workers. The requirement, therefore, is to ensure safety as far as is reasonably practicable. But how can employers demonstrate they are managing hazards and risks to be as low as reasonably practicable (ALARP)? It is not possible to define ALARP in purely objective and absolute terms. There will always be a need for experienced judgment and subjective opinion, and hence there will always be potential for debate. Therefore, employers need guidance, in the form of guidelines and codes of good practice. In practice, the regulatory regime tends to revert to a flexible form of prescription. Public-authority involvement concerning prescriptive regulation has increased over the past decades and Oh (2002) points out that this involvement will continue to grow in the future. For example, the *domino effect* article of the Seveso II Directive in Europe (Council Directive 2003/105/EC, 2003) imposes an "involvement" of neighboring companies with each other, by demanding information exchange on the potential of multi-plant accidents. At present, the requirements and the depth of such cross-company information exchange are still open to debate. Security will undoubtedly also play an increasingly important role in the public and private debate in protecting critical assets against knock-on events by means of inter-company collaboration.

The most optimal way to install a multi-plant safety and security culture is that captains of industry are convinced of its added value to the cluster. Section 2.7 indicates the huge hypothetical financial benefits from jointly installing such a multi-plant initiative, which might serve as an important incentive for the decision-making process. Furthermore, Chapter 9 discusses a game-theoretical approach that can be used to convince top management of neighboring plants of the financial benefits of cross-plant collaboration.

By establishing a multi-plant safety and security culture, captains of the chemical industry would truly decrease the likelihood of a multi-plant accidental domino effect or a multi-plant terrorist attack. They would also anticipate *ad hoc* legislation. Figure 3.1 illustrates the tendency of major accidents to force politicians to take political actions *ad hoc*, presenting a chronological overview of a non-exhaustive number of significant major chemical accidents worldwide, on the one hand and the chronological developments in a non-exhaustive number of European and US safety regulations on the other hand.

All political attention comes to focus on the accident that has just happened, while a needed pro-active broad political view of the chemical accident prevention issue as a whole is overlooked. The reason for this re-active political behavior is that politicians and company policy makers have limited imaginative power to fully understand the probabilities of accident estimates. Moreover, people are risk averse for gains, but risk taking for losses. This phenomenon causing *ad hoc* prevention legislation is one of a number of biases in information processing that occurs when people make choices. Originally described by two psychologists, Kahneman and Tversky (1984), the work has had a considerable impact on economic models of choice. Tversky and Kahneman discovered there is a strong tendency for individuals to be what they called "risk averse" for gains, but "risk taking" for losses. The choice experiment originally formulated by Tversky and Kahneman has been carried out many times, in many different situations, and the results are very robust. Whether the decision is presented as a loss or a gain will influence what decision is taken. People prefer to hang onto their gains, but gamble with their losses. In other words losses and gains are not equally balanced in decision making. Certain gains are weighted far more heavily than probable losses. As Sutherland *et al.* (2000) point out, this can be easily seen in the balancing of

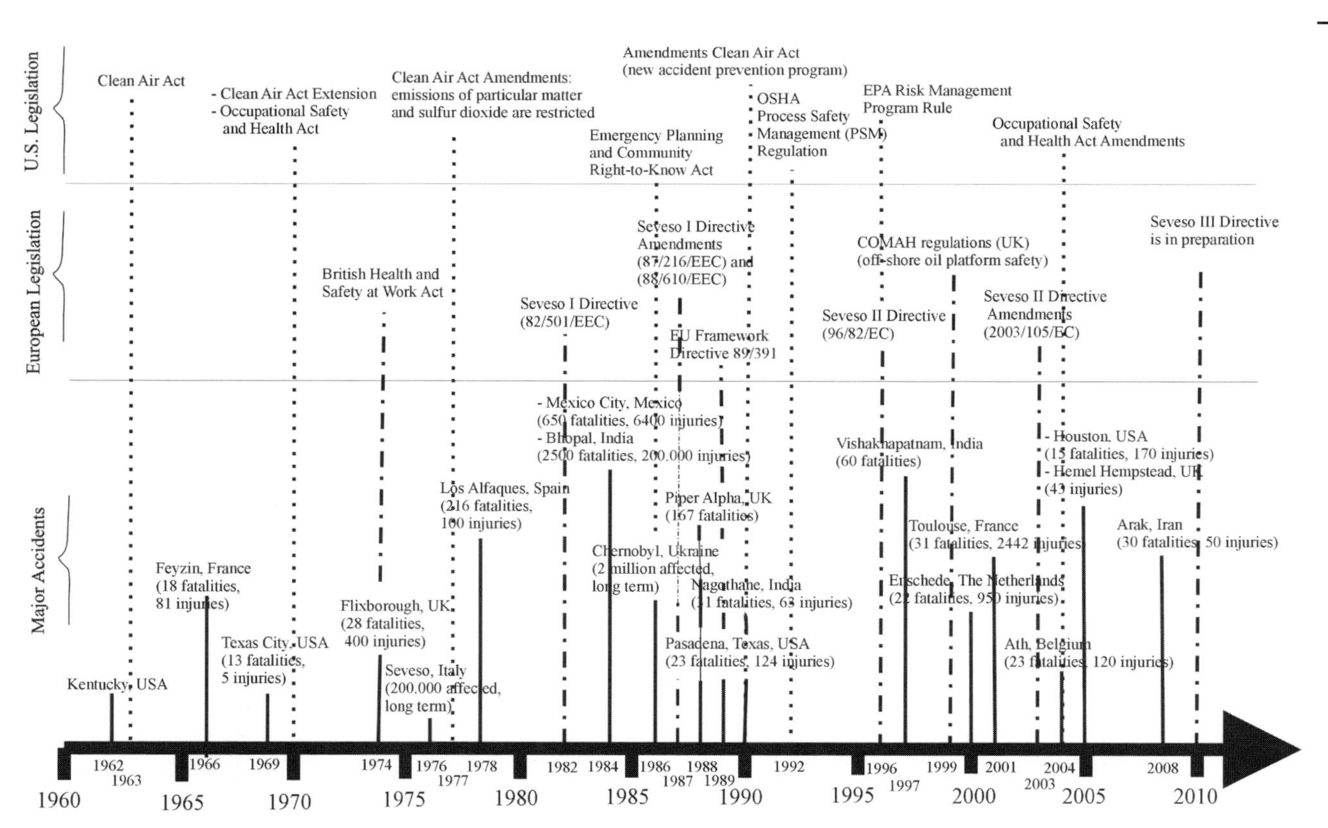

Figure 3.1 A non-exhaustive number of major accidents and their relation with some European and US safety legislation.

"production" and "safety". Ignoring some safety aspects will lead to almost certain increases in production. Applying safety measures for highly improbable accidents will lead to additional costs. Hence, it takes a much larger, and more probable, loss to tip the prevention management decisions in favor of safety with respect to external domino risks.

Furthermore, only since the events of September 11, 2001, have governments and experts in the field truly recognized the potential threat that chemical facilities pose. Many plants house toxic chemicals that could become airborne and drift to surrounding (populated) areas or be used to create weapons capable of causing substantial harm. Chemical clusters could even be largely destroyed by deliberately induced domino effects, causing many fatalities and injuries and considerable – and often irreversible – economic damage.

No (holistic) European Directive establishing minimum safety and security standards for integrated chemical clusters exists, nor is there any US legislation on the subject. Therefore, to advance multi-plant safety and security, individual plant managers should be convinced of the huge (social and economic) hypothetical benefits of voluntary installing a well-considered safety and security culture within their industrial area or within a part of a larger industrial area. Considering current industrial safety and security practices in chemical plants, this book deals with guidelines on how to set up such a multi-plant safety and security culture.

3.3 The Present State-of-The-Art to Deal with Safety and Security Risks

3.3.1 A Plant-Safety Culture

The chemical industry possesses an unusual degree of technological uniformity worldwide and it is widely recognized that the industry is very safety driven. Yet the probability of becoming involved in a major accident with at least one fatality varies a lot between companies. While factors such as national and company resources will play their part, differences in plant-safety cultures are likely to contribute the lion's share to the variation.

A plant-safety culture can be engineered by identifying and fabricating its essential components and then assembling them into a working whole. Thus, acquiring a safety culture is a process of collective learning, made up of a number of codes of good practice. The goal of a plant-safety culture is to encourage plant personnel to think and manage things in such a way that enhanced safety becomes a natural by-product. The following definition was given by the UK's HSE in 2005:

> The safety culture of an organization is the product of individual and group values, attitudes, competencies and patterns of behavior that determine the

> commitment to, and the style and proficiency of, an organization's health and safety programmes. Organizations with a positive safety culture are characterized by communications founded on mutual trust, by shared perceptions of the importance of safety, and by confidence in the efficacy of preventive measures.

In other words, a plant-safety culture is the effort with which all organizational members direct their attention and actions towards improving safety on a daily basis within the company. Although a culture cannot easily be created, as explained by Schein (1990), the engineering or the enhancement of a safety culture relies on the deliberate manipulation of various organizational characteristics thought to impact upon safety-management practices. An approach for streamlining these manipulations must be goal directed. Goal-setting theory may also serve as the necessary scientific utility for the safety-culture concept, as becomes apparent when the purposes of the safety-culture definition are examined. These include producing behavioral norms, reducing accidents and injuries, ensuring that safety issues receive the attention warranted by their significance, ensuring that organizational members share the same ideas and beliefs about risks, increasing people's commitment to safety, and determining the style and proficiency of an organization's health and safety programmes. Hence, the creation of a safety culture can be achieved by dividing the task into more manageable sub-goals that are in themselves challenging and difficult (e.g., conducting risk-assessment procedures, getting operators to "think safety", convincing senior managers to use state-of-the-art guidelines, *etc.*). To manage these sub-goals efficiently, the current state-of-the-art concept of a "plant-safety culture" can be analytically described as a composite of people, procedures and technology using systematic engineering and management techniques to keep all systems within a given environment safe throughout their life cycles. Figure 3.2 represents the different parts composing a plant-safety culture.

For each of these building blocks (i.e. people, procedures and technology) needed by an organization to efficiently perform its safety tasks, generally accepted concepts exist within the chemical industry. To deal with the procedural needs, so-called "plant-safety-management systems" are elaborated by the individual companies; to address the technology building block, company-specific tools, semi-quantitative risk analyses and quantitative risk-analysis software packages are used. The company personnel responsible for enhancing all aspects of company safety (providing safety training, *etc.*) is referred to as "plant-safety management".

Although nowadays many chemical companies have their own plant-safety culture, plant-safety managers recognize that such an individual culture approach does not lead to the level of cross-plant accident prevention needed for the complex situation of a chemical cluster (Reniers *et al.*, 2005a). Multi-plant accidents can only truly efficiently and effectively be prevented through systemic and guided safety collaboration between chemical companies located next to one another and hence forming an industrial cluster.

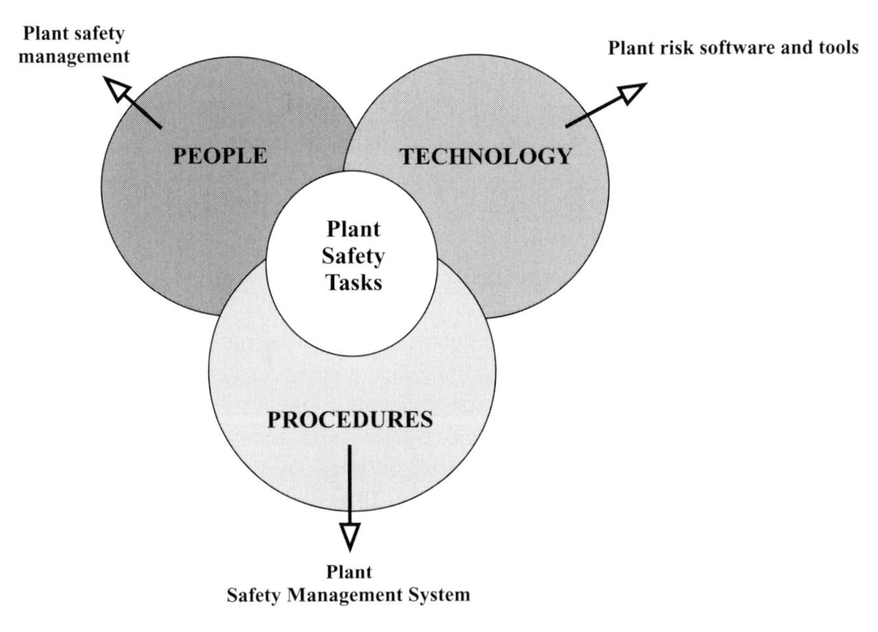

Figure 3.2 People, procedures and technology performing plant-safety tasks.

3.3.2
A Plant Operator Security Plan

The goal of a plant Operator Security Plan (OSP) is to evaluate whether an adequate level of security is provided at a chemical company. The OSP reduces security risks associated with malicious acts and also provides for adequate mitigation should such an act occur. An OSP thus shares many commonalities with plant-safety-management systems. Although for example, identifying and managing the hazards of chemical releases as well as elaborating emergency-response plans are common activities, clearly, safety programs do not explicitly address evil-minded acts such as contamination of products designed to cause harm to the public, theft, or the diversion of hazardous materials. Therefore, rather than devising completely separate and diverse programs to deal with both intentional and accidental acts involving hazardous substances, in many chemical enterprises parts of these programs are integrated. Nonetheless, since an essential part of the security plan within a chemical company is also quite different from plant safety activities (such as, e.g., acccss control, drug and alcohol use, weapons carried by employees, information protection, suspicious packages, terrorist threats, *etc.*), managing security activities in a chemical plant can be considered on the same management level as managing safety tasks. Bearing the latter in mind, an OSP should be viewed within the larger context of the three building blocks shaping the plant-safety culture.

Each of the three building blocks for shaping a safety culture deals with security issues in addition to safety issues. As such, management of all safety- and security-related activities by the employees of a chemical company is to be indicated as the "plant safety *and security* culture".[1] This terminology indicates the operator security plan, the manner in which employees think in terms of security (attitudes, values, and competencies), the various technological tools and the organizational measures to enhance plant security. Obviously, these all form an integral part of the safety and security culture of the company.

3.3.3
Cooperative Strategies in Chemical Clusters

As Fortis and Maggioni (Curzio and Fortis, 2002) state, firms decide to settle in a cluster on the basis of the expected profitability of being located there. This profitability depends on geographical and agglomeration benefits, obtained as the difference between gross location-related benefits and costs. As the number of corporations located in an industrial cluster increases, gross benefits increase due to productive specialization, scientific, technical and economic spillovers, reduction in both transport and transaction costs, increases in the quality of the local pool of skilled labor force, *etc.* This observation also explains why chemical plants form chemical clusters. However, in the case of chemical enterprises, clustering not only implies profit opportunities and economic benefits of scale. A chemical cluster has a very high responsibility towards maintaining safety and security standards in the urban surroundings as well. Each additional chemical plant entering a chemical cluster might decrease the average safety and security standing of the area.

Companies in chemical clusters are thus not merely linked via technological spillovers, logistics advantages, and so on. They are related through the responsibility of gaining and sustaining safety and security standards in the entire cluster as well.

To maximize the clustering gross benefits, chemical organizations have a long tradition of collaborating on many different fronts. Their cooperative strategies offer significant advantages for plants that are lacking in particular competencies or resources to secure these through links with other firms possessing complementary skills or assets. They also offer opportunities for mutual synergy and learning.

However, as Child *et al.* (2005) explain, the organizational cultures can have a significant impact upon the implementation of cooperative strategies. Cultures can create serious barriers to collaboration between organizations, and yet, at the same time, the knowledge embodied in cultures can provide a valuable resource

1) It is important to explicitly mention all security topics to form an integral part of the company's safety culture. Therefore, in this book a plant's traditionally called "safety culture" is called the "plant's safety *and security* culture".

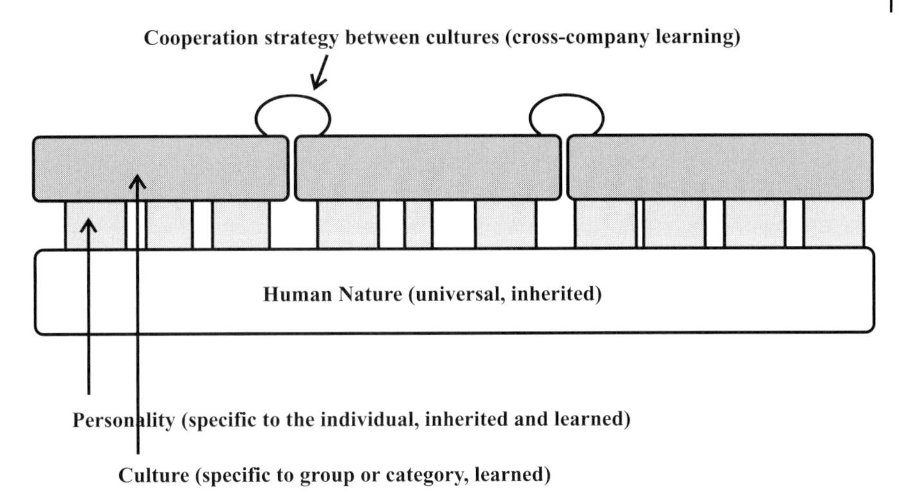

Figure 3.3 Culture contrasted with other sources of human thinking and behavior (Source: based on Child *et al.* 2005).

for cooperation strategies. To better understand the latter reflection, Figure 3.3 contrasts *culture* with two other important influences that shape the ways in which people think and behave, namely *personality* and *human nature*.

It is widely accepted that culture is something that is learned. It is not inherited like human nature or, at least partially, personality. This has a very important practical implication for cooperation between the members of organizations coming from different cultural environments, as it means that despite the way a certain company culture is absorbed by a plant's employees, it may always be possible to learn further cultural attributes through experience or training. In terms of safety and security topics in particular the latter observation is interesting. Safety and security matters are often subject to a high degree of confidentiality. Therefore, safety and security cooperation strategies are difficult to establish. However, since clustered chemical corporations are bonded by the responsibility to keep the industrial area *as a whole* as safe and as secure as possible, individual plants situated next to one another should develop a safety and security cooperation strategy, bringing diverse safety and security cultures together. Shaping such a site-integrated safety and security culture can lead to significant advantages for the chemical cluster as a direct result of cross-organizational learning. Avoiding incidents and accidents and their associated direct and indirect costs can lead to substantial benefits (especially in case of averting major accidents such as external domino effects, see Section 2.5). Although the potential benefits are obvious, developing a multi-plant safety and security culture could not be described as an easy task. It is obvious that cooperation between chemical plants is a basic premise to attain a multi-plant safety and security culture. How to enhance cooperation between chemical plants is discussed in the next section.

3.3.4 Enhancing Collaboration in Chemical Multi-Plant Areas

Cooperation and competition provide alternative or simultaneous paths to success. Hence, in business, as in nature, decision makers must be aware that competing and collaborating are equally valid aspects of corporate strategy. Although cooperative arrangements within many industries are well known and often successful and appreciated, further optimization of these arrangements is often possible. By augmenting collaborative agreements and relationships and by linking up with other firms on the same level of the market, a company may enjoy options otherwise unavailable to it, such as better access to markets, pooling or swapping of technologies and production volumes, access to specialized competencies, lower risk of R&D, enjoying larger economies of scale, benefiting from economies of scope, *etc.* (De Man *et al.*, 2000; Contractor and Lorange, 2002). In this section the pro-active extension of collaborative relationships are examined by analyzing the chemical companies' decision-makers' perceptions on "successful collaboration". Competition-based perceptions and requirements of managers deciding on collaborative agreements are thus (indirectly) taken into account.

If companies find themselves cooperating on issues and domains other than safety and security, they are likely to be less hesitant to also collaborate on the latter topics (which might in some cases be perceived as highly confidential). Therefore, elaborating such a framework would appear to be useful for multi-plant safety and security collaboration as well.

In their study, Reniers *et al.* (2009e) surveyed shippers from the chemical industry. Both cooperators and non-cooperators were identified within the survey, allowing the evaluations of both types of respondents to be compared. The research concluded that the concerns of those who do not cooperate are indeed supported by empirical data from those cooperating horizontally. Summarizing, non-cooperators indeed have a realistic perception of cooperation in general and of partnerships in particular and seem to require no additional information or knowledge on these topics. Hence, it was possible to use the survey data from both cooperators and non-cooperators to further analyze both groups' perceptions on collaborative arrangements and to draw conclusions.

A framework for advancing and stimulating horizontal collaboration in the chemical industry (ASC) was developed based on insights from the literature and empirical data on the search process for partners. The framework enables researchers and industrialists to gain better insights into the success and failure of horizontal collaborative relationships. The framework builds on a collection of *collaboration drivers* and *partner features* applied collectively to reveal theoretical "collaboration constructs" for horizontal collaboration which cannot be assigned directly (De Vellis, 1991). Based on their nature, collaboration drivers and partner features can be classified into "hard factors", "soft factors" and "independent factors". Hard issues typically concern a more rational, calculative, concrete, perceptible, tangible nature, whereas soft issues are associated with a non-rational, cultural, social, non-measurable, abstract, imperceptible, intangible disposition.

Table 3.1 Classification of collaboration drivers and partner features.

Hard factors	Soft factors	Independent factors
Financial opportunities offered	Trust	External flexibility
Service level offered	External willingness to collaborate	Level of supplementarity/ complementarity
Market position – relative bargaining power	Cultural fit between companies	External innovation potential
Necessary investments for collaboration	Openness between companies	Internal stakeholder support and commitment
Benchmark results concerning potential partner		Former partnerships and experiences
External financial position		
External knowledge		

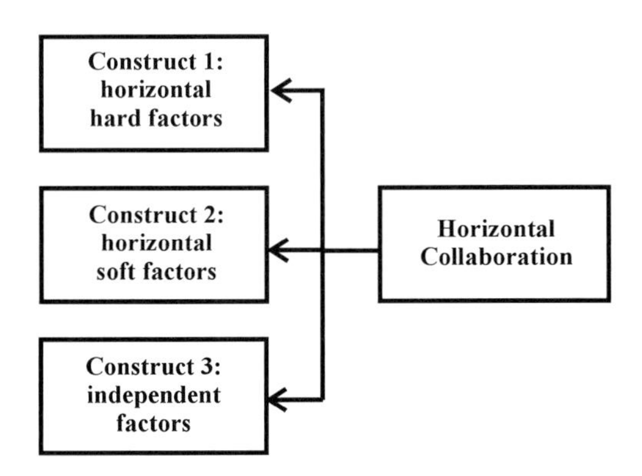

Figure 3.4 Conceptual model to develop the ASC framework.

General factors cannot be classified into one of the former types. A collection of measuring items (i.e. collaboration drivers and partner features) was employed to indirectly empirically validate the extent to which the different kind of factors contribute to horizontal collaboration success.

A total of 16 measurement items were identified by the literature and in-depth interviews: 4 collaboration drivers and 12 partner features. These items were further categorized by expert opinion into hard factors, soft factors and independent factors. Table 3.1 presents the classification results.

The main objective in this section is the development and empirical validation of a collaboration enhancement framework capturing horizontal collaboration. Figure 3.4 illustrates the variables (also referred to in the scientific literature as latent constructs) and the theoretical model.

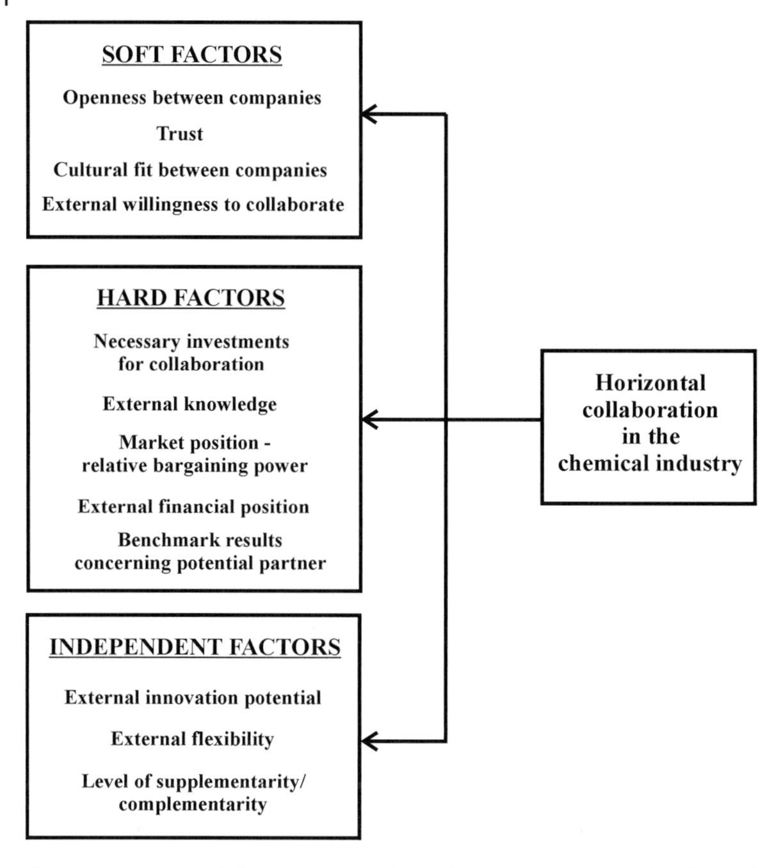

Figure 3.5 Framework for advancing and stimulating cooperation in the chemical industry.

To develop the constructs and the collaboration models the well-established methodology of Churchill (1979) as refined and extended by Gerbing and Anderson (1988) was followed.

Since the formation of the theoretical links was undertaken *a priori*, confirmatory factor analysis is the suggested method for empirical validation (e.g., Swafford *et al.*, 2006). The resulting ASC framework is depicted in Figure 3.5.

From the framework illustrated in Figure 3.5 it can be seen that only 12 out of the 16 originally identified parameters have a significant impact on horizontal collaboration success perception in the chemical industry. Apparently, all 4 initially identified "soft factors" are perceived as essential.

Chemical companies belonging to a chemical cluster may employ the ASC framework to consider and to investigate the identified features and drivers for the possibility of a desired type of collaborative understanding for forming a multi-plant area (eventually being part of a larger chemical cluster). By doing so, the likelihood of a stable relationship is enhanced and the initiated collaboration may grow towards a multi-plant safety and security culture.

3.4
Coping with the Future: Developing a Multi-Plant-Safety and -Security Culture

Cooperative strategies bring together people from different organizations and place them in a working relationship. The organizations from which the people come will each have developed their own distinctive safety and security cultures. These cultures embody shared attitudes and norms of safety and security behavior. They usually encourage people to regard their organization as different from, and often as superior to, other organizations, and therefore to hold on to their ways of doing things. Thus, as already indicated, given the vested interests of various stakeholders and their different approaches to safety and security problems, elaborating an integrating approach for enhancing multi-plant safety and multi-plant security in the chemical industry is a complex and delicate exercise. However, the success of many businesses arises from managers viewing the industry and its interacting activities as a whole, thereby demonstrating the ability to foresee future developments. For example, evolution with respect to safety management in chemical clusters is moving towards plant-safety managers integrating their knowledge, actions and communications to provide knowledgeable foresight concerning complex multi-plant-related safety and precaution issues.

The current non-existence of plant safety and security managers merging the various safety and security approaches and risk-analysis techniques used in the different chemical plants, relating them to each other, and using this relationship to enhance multi-plant safety and multi-plant security (e.g., by more efficiently preventing external domino accidents) is due to two main factors. First, individual plant-safety management and plant-security management is often unwilling to cooperate due to data confidentiality and corporate policy. Second, recommendations and guidelines for elaborating a multi-plant system safety and security culture simply do not exist.

The first of these factors can be remedied by forcing the new ideas and guidelines proposed in this book into practice, as the government has done on nuclear safety and other areas of public concern. However, the decision by chemical industries to use a multi-plant safety and security culture approach should be to achieve better safety and to be better protected against intentional mischief, not just to comply. A compliance-oriented approach has two major shortcomings. First, codes, standards, and regulations are, by and large, political documents. As such, they are frequently the result of compromise and represent minimum acceptable levels of performance. Even though full compliance is sometimes a challenging goal, a safety program as well as a security program achieving only such government-driven (safety and security) documents meet only minimum standards. A second shortcoming of safety programs or security programs based only on compliance is that they tend to be reactive, as was also proven in Section 3.2. It therefore becomes clear that there always exists a gap between compliance-driven safety and security on the one hand and optimum safety and security on the other, and legislation tends to be inadequate for leading-edge safety and security activities. It can only be hoped that regulations for enhancing safety issues

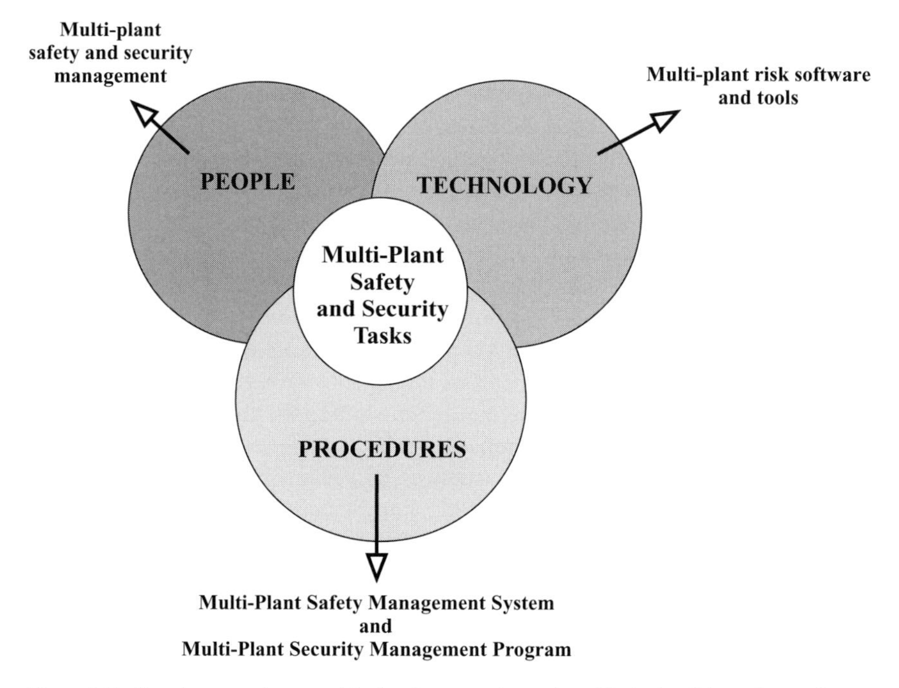

Figure 3.6 People, procedures and technology carrying out multi-plant safety and security tasks.

and security topics in clusters composed of chemical installations will not come about as a result of for example, an intentionally designed multi-plant domino effect.

To counter the non-existence of guidelines dealing with multi-plant safety and security, this book formulates recommendations for developing a holistic approach to tackle multi-plant system safety and security topics. Moreover, recommendations are offered for minimizing reciprocal distrust between individual plants. A resulting multi-plant safety and security culture (M-PSSC) emerges as a solution to long-range chemical cluster safety and security success. The concept used for elaborating such a M-PSSC can be regarded in an analogous way to that of a plant safety and security culture, as illustrated in Figure 3.6.

The three building blocks necessary to efficiently perform multi-plant safety and security tasks are each thoroughly elaborated further in this book. Research-based guidelines and recommendations were developed for addressing the concepts of each of the building blocks.

It is recognized that the safe functioning of an organization depends on its safety-management system (being part of the plant's safety culture). The organization's security hinges on its security-management program (being part of the operator security plan). These statements can easily be broadened to a cluster of organizations. Any plant-safety- -management system/plant-security- -management program may be dependent on a safety-management system/security-management program developed for a larger entity such as a group or cluster of establishments. The safe/secure operation of a group of establishments requires guidelines for the implementation of a system of structures, responsibilities and procedures, with the appropriate resources and technological solutions available. These guideline procedures, called the "multi-plant-safety management system" (M-PSMS)/the "multi-plant-security management program" (M-PSMP) as proposed in this book are discussed in Chapter 4. Chapter 5 investigates people that have to deal with multi-plant safety, respectively multi-plant security, known collectively as multi-plant-safety management, respectively multi-plant-security management. The chapter proposes new frameworks for enhancing cooperation between plant-safety managers and between plant security personnel. By doing so, an industrial area can be more successfully investigated in terms of potential multi-plant domino effect hazards. Chapter 6 deals with the engineering of adequate technology (e.g., a decision-support system for more effectively abating cascade-effect risks in an industrial area by prioritizing the installations based on the danger they present in triggering external domino effects).

The three resulting multi-plant building blocks are, by and large, entangled. For example, the data needed as input for the multi-plant decision-support system and other information on technological safety and security issues are given by the individual plant safety and security managers as stipulated by the M-PSMS and the M-PSMP. The input for the frameworks proposed for ameliorating cross-company cooperation is provided by the multi-plant decision-support system. The quality and the continuous improvement of the M-PSMS and the M-PSMP is partly a consequence of the solid functioning of the multi-plant frameworks. Moreover, all these relationships are bidirectional, and thus characterized by a necessary feedback of information.

Another way of looking at the M-PSSC is by using an object-oriented approach. A three-dimensional representation (see Figure 3.7) can be used to illustrate the different objects or building blocks putting together the M-PSSC model. The modeling method can be based on abstracting the building blocks to arrange the multi-plant safety and security goals into three dimensions: operational, tactical, and strategic.

To construct the M-PSSC model, these three views have to be unambiguously identified. The strategic dimension is characterized by structuring the subcomponents of the model and by streamlining the interconnection among the different building blocks. The tactical dimension is characterized by investigating which multi-plant accidents can happen in the cluster and what measures can be taken to prevent them. The operational dimension provides the tools, algorithms or computations for determining where (at which locations) multi-plant accidents can happen and based on information gathered from all plants participating to the multi-plant initiative. The advantages of such object-oriented modeling are (i) the relative independence of the different building blocks, leading to relative simple

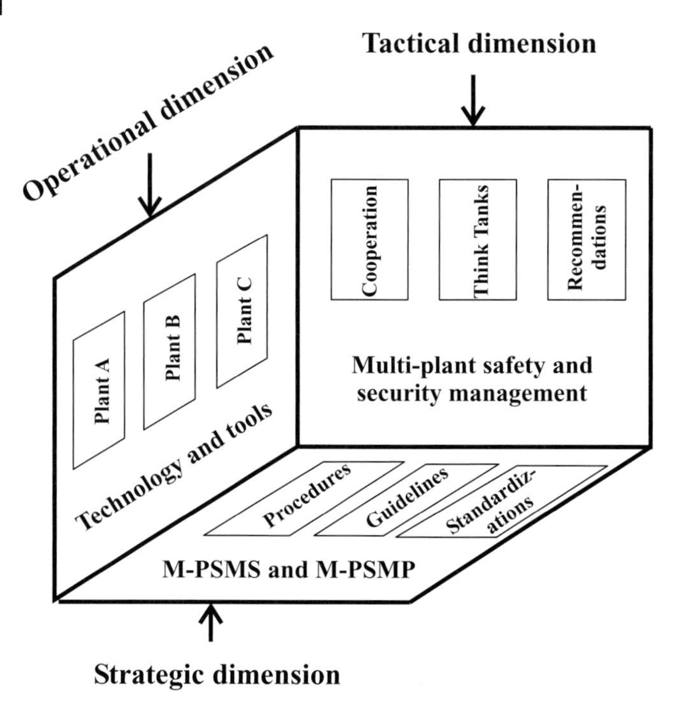

Figure 3.7 Object-oriented representation of the M-PSSC model.

optimization of the building blocks if needed, and (ii) the ease of reuse of (parts of) the building blocks in case of different chemical multi-plant areas.

To illustrate the use of such a 3D representation, a three-dimensional point situated somewhere in the "multi-plant safety and security space" can be viewed upon as a multi-plant safety and security issue. The issue is unambiguously defined by its three dimensions (i.e. operational, tactical, and strategic). Guidelines concerning the topic can be found in the strategic dimension, whereas the concrete continuous improvement and evaluation of these guidelines and of the issue itself are contained in the tactical dimension. The operational dimension investigates the relationship of the issue with other issues in the multi-plant safety and security space and calculates its relative danger towards multi-plant safety and/or multi-plant security. In this way, the topic is positioned within the multi-plant danger space, the latter being dually connected with the multi-plant safety and security space.

To help companies within a chemical cluster organize these three building blocks, a multi-plant platform should be elaborated within the cluster. Designing such a platform is not at all an easy task. The platform should not be one concrete organizational structure, but it should be a conglomeration of many different and diverging organizational structures. It should be an organizational platform wherein all safety and security achievements and results (from the different companies composing the platform) are reported and processed to continuously improve individual plants' safety and security as well as multi-plant's safety and security. In Chapter 4, the notion of a multi-plant platform or a so-called multi-plant council is further elaborated.

3.5
Summary and Conclusions

This chapter has identified the main components for establishing an integrated multi-plant safety and security culture, which is essential to move to the next generation of dealing with safety and security in the chemical industry. Within a site environment, procedures, people and technology are considered to elaborate an integrated system to manage all safety and security aspects within an industrial area throughout its lifecycle. The building blocks that have to be well worked out to this end, are a multi-plant safety and security-management system, multi-plant management collaboration frameworks to continuously improve safety and security within the cluster as well as within individual plants (based on shared knowledge and joint know-how), and multi-plant technology and tools to deal with the technical aspects of preventing multi-plant accidents.

The multi-plant safety and security culture modeling should be object oriented and hence the possibility exists to define multi-plant safety topics as well as multi-plant security issues as associated with each model building block. In this way, individual plants have the opportunity to implement multi-plant strategic, tactical and operational codes of good practice, recommendations, directives, *etc.*, in an adjusted way taking their specific enterprise context into account.

Although chemical enterprises individually already address safety and security issues, they do this from a multi-plant perspective in an uncoordinated way which inevitably leads to a less-than-optimal use of resources and experience. In a number of cases, individual safety and security measures might fail to meet the highest standards of a chemical cluster and they might not be sufficient to guarantee multi-plant safety and security. Establishing a type of multi-plant platform or so-called multi-plant council within a chemical multi-plant area will streamline the entire process of setting up and organizing the three required building blocks and it will without doubt lead to enhanced preventive multi-plant safety and security.

4
A Multi-Plant Safety and Security Culture – The Procedures: Establishing a Multi-Plant Safety and Security Management System

4.1
Introduction

There is need for guidance and advice for organizing and implementing concepts to bring multi-plant safety and multi-plant security to the next level. Requirements and recommendations of existing plant-safety-management system and plant-security management program codes of good practice should be analyzed in relation to the needs of multi-plant chemical safety and security. Comprehensive guidelines are to be established for gradually standardizing plant-safety management systems and plant-security management programs through the design, the development and the installation of a multi-plant-safety management system and multi-plant-security management program within a group of chemical companies. A multi-plant organization framework and a scheme for continuously improving single plant and multi-plant safety and security management via communication and cooperation at the plant department level as well as at the multi-plant level is required. Therefore, in this chapter, recommendations are given to elaborate the first cornerstone of a multi-plant-safety and -security culture, that is, how to engineer multi-plant-safety and -security-management procedures.

During the past decades, systematic health and safety management has become increasingly popular among both regulators and employers as a central strategy to reduce accidents and illness at work (Frick *et al.*, 2000). Safety management aims to do this by using managerial processes to detect and reduce workplace hazards. Security management became a major political issue as well as an important business issue after September 11, 2001 (Alexander, 2004). Security-management programs are being drafted, implemented and evaluated to map, to analyze and to counter potential unlawful acts, perpetrators, objectives, intended outcomes and motivations, targets, and *modus operandi*.

The chemical industry is no exception to this rule. Since the safety or the security of complex well-defended systems in a chemical plant is dependent upon a variety of contributory human, technological and organizational factors, managing health and safety and indeed security in such companies is not an easy activity. Moreover, as already mentioned, in the (petro-)chemical industry economies of scale, environmental factors, social motives and legal requirements often force companies

Multi-Plant Safety and Security Management in the Chemical and Process Industries. G.L.L. Reniers

ISBN: 978-3-527-32551-1

to "cluster". Also, the incidence and the severity of accidental accidents tend to increase (Khan and Abbasi, 1999), as do the vulnerability of the industrial area and the potential threat from terrorism (Alexander, 2004).

Three kinds of accidents can be distinguished: those that happen to individuals, those that happen to organizations and those that happen to clusters of organizations.[1] Reason (1997) indicates that individual accidents are by far the largest in number and that organizational accidents are comparatively rare, although often much more serious. However, Reason (1997) does not mention the potentially most catastrophic type of accidents, that is multi-plant accidents, probably because of the extremely low rate at which they occur. Regrettably, such accidents do occur, sometimes with disastrous consequences (see also Table 2.4). Multi-plant accidents can be related to linked production and/or linked delivery of services, as well as to cross-company (or external) knock-on effects. The first type of multi-plant accidents implies cluster-related safety problems might occur as a result of divided responsibilities. Such accidents generally do not have catastrophic consequences; however, the resulting economic losses may be substantial. The second type of multi-plant accident is the result of external domino effects and can be disastrous in terms of the loss of lives. Available reports on domino effects (e.g., Lees, 1996; Pietersen, 1986, 1990) describe multiple accident events that have taken place in the past. The worst such accident – in terms of death toll – occurred in Mexico City on November 19, 1984. The accident claimed 650 lives and injured approximately 6400 people (see Table 2.4). It was an external domino effect involving three companies: the PEMEX plant (where the accident originated), the Unigas plant, and the Gasomatico plant.

Considering multi-plant accidents, it is obvious that both process safety (and security) and occupational safety (and security) should be dealt with to obtain multi-plant safety improvement.[2] To prevent major multi-plant accidents, however, preventive efforts in a cluster context should be stressed on process safety (and security). Although process safety accidents are often related to complex failure scenarios including both process hazards and occupational hazards, the former (process-related) hazards represent a larger contribution to the possible complicated mechanisms of such accidents. It should be noted, however, that occupational safety (and security) plays a very important (indirect) contributory role to sound and solid process safety (and security) management.

The emphasis of enhancing safety (and security) in a chemical multi-plant context lies in the prevention of external domino effects. To identify which chemical plants pose an escalation risk or threat to one another and to what extent, software and instruments are internationally being developed. *Software examples* include Aidram-Cargo (Switzerland), ARIPAR (Italy), Charm (USA), DominoXL (Belgium), Flacs (Sweden), Fred/Sheperd (UK), SAVE (The Netherlands), THESIS (USA), *etc.* The interested reader is referred to Reniers *et al.* (2006). These tools

1) Consequences of individual accidents mainly relate to individual employees (e.g., most work-related accidents), the outcome of organizational accidents affects the entire organization (e.g., internal domino effects), and cluster-related accidents have an impact upon several chemical plants within an industrial area (e.g., external domino effects).

2) Process safety hazards are those arising from the processing activities of a chemical company. Occupational safety hazards affect individuals but have generally a minor impact on the processing activities of a plant.

are either used to identify the potential for domino effects and/or the resulting scenarios of major accidents. *Instrument examples* include the Belgian domino methodology (Delvosalle, 1998), the domino risk identification and evaluation instrument (IDE) that has been elaborated on request of the Dutch authorities (RIVM, 2003), the MICADO instrument that was developed in France (INERIS, 2002), the ARIPAR instrument developed in Italy (Spadoni *et al.*, 2000, 2003), *etc.*

Three main justifications can be cited for drafting both a safety-management system and a security-management program on a multi-plant level.

First, the literature, for example, Lees (1996) or Ioannidis *et al.* (1999), indicates that the prevention of large-scale industrial accidents, such as domino accidents, should be based on reliable risk-assessment studies. However, no guidelines exist to advise multi-plant managing centers handling rational and quantitative information on how to estimate the consequences of multi-plant major accidents and how to propose preventive action (related to safety risks as well as security risks). Therefore, it seems logical that it should not only be individual chemical plants that elaborate managerial procedures and guidelines for preventing major incidents. The development of a management document on a multi-plant level dealing with multi-plant safety and security issues such as the prevention and mitigation of cross-company accidents should also be contemplated.

Second, nowadays, chemical plants wishing to obtain certain permits must agree to contract external organizations for their single-plant escalation-effects studies, indicating the danger of the plant towards its environment. However, multi-plant studies are not compulsory, and as a result are often not included. Given the complex nature of multi-plant risks and the involvement of different chemical enterprises, it is very difficult to obtain all the necessary confidential information to assess multi-plant events. Moreover, there are some difficult questions to answer before the industry will be convinced of the need to enhance multi-plant or external domino-prevention cooperation. There is the question as to who will perform the study, how will it be performed, and how will the implications of the study and the proposed preventive measures be executed. Personnel with specific expertise in terms of the cluster, but independent of the cluster plants, would be able to handle confidential information and to provide suggestions.

Third, to the best of the author's knowledge there has been no attempt to formulate an integrating approach for the provision of legislative suggestions on swapping domino-effects information. Management guidelines on the subject have not yet been formulated.

4.2
Managing Safety, Quality, Environment, and Security

4.2.1
Introduction

Management systems for the safe operation of chemical plants require a system of structures, responsibilities, procedures, and the availability of appropriate

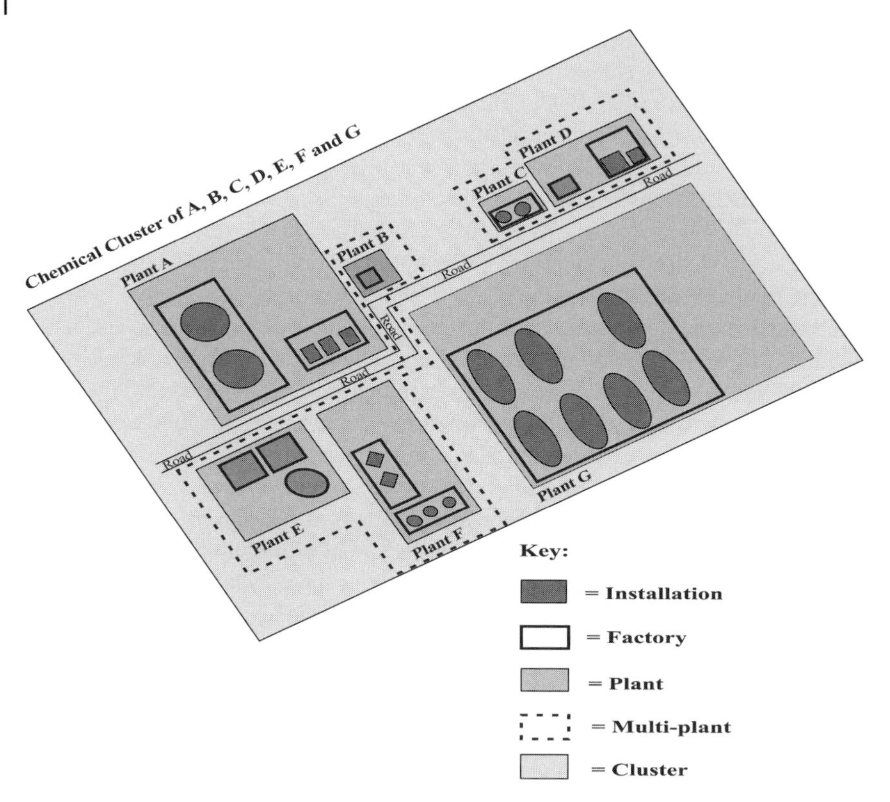

Figure 4.1 Different levels within chemical clusters: installation, factory, plant, multi-plant and cluster.

resources and technological know-how. Safety in a chemical surrounding can be managed at four different levels: at installation, at factory (or business unit), at plant (or company, enterprise), at multi-plant and at cluster level. Figure 4.1 illustrates the difference between the latter terms. Plant *B*, for example, consists of only one installation within a single factory. Plant *B* forms a multi-plant area with plant *E* and plant *F*, all three plants participating in the same multi-plant safety and security initiative. Plant *A* and plant *G* choose not to collaborate intensively on a supra-plant level, whereas plants *C* and *D* form their own small-scale multi-plant safety and security initiative.

Factory-level safety includes topics such as working procedures, work packages, installation-specific training, personal protective equipment, quality inspection, *etc.* Plant-level safety includes defining acceptable safety risk levels and documenting plant-specific guidelines for implementing and achieving these levels for every facility situated on the premises of the plant. It is current industrial practice to draft a plant-safety management system (PSMS) to meet these goals (see also Chapter 3). Multi-plant level safety-related topics include defining multi-plant safety standards, defining safety cooperation levels, defining acceptable multi-plant risks, joint

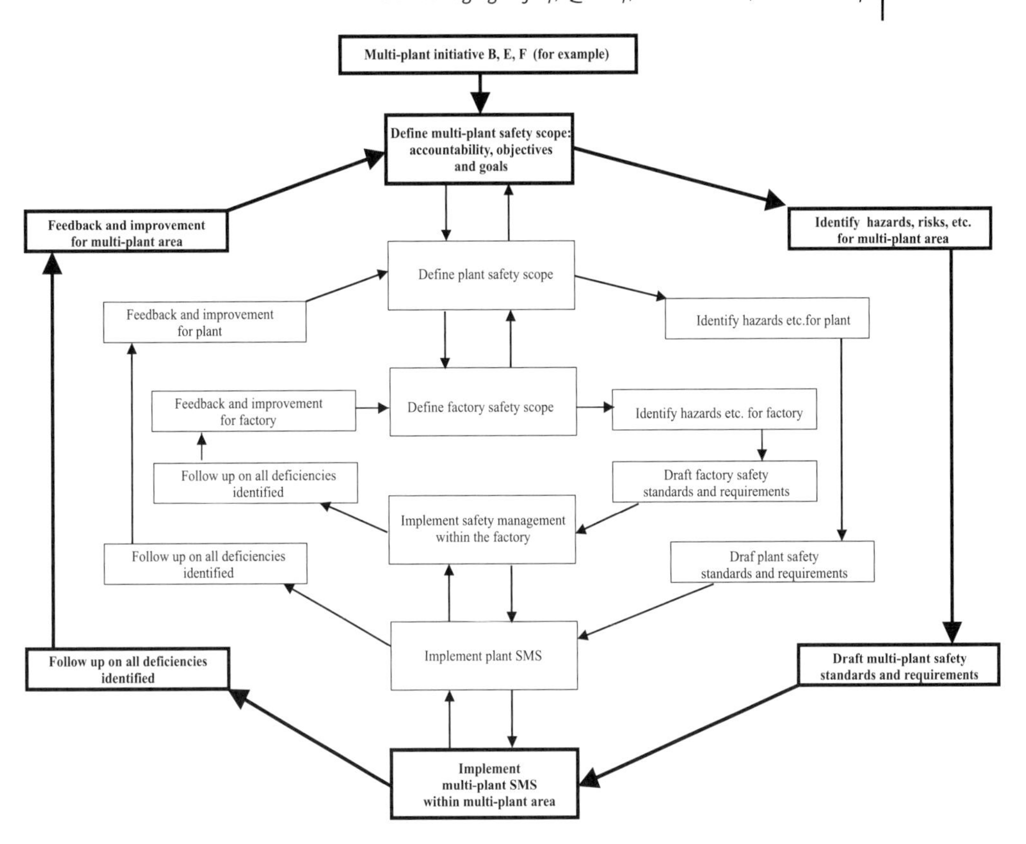

Figure 4.2 Loop safety structure on factory, plant, and multi-plant levels.

workforce planning in the event of multi-plant accidents, joint emergency planning, *etc.* The latter multi-plant-related topics can be documented in a type of safety-management system dealing with these issues. In the chemical industry, at factory level and at plant level, safety documents, guidelines and instructions, technical as well as meta-technical, are usually very well elaborated. An adapted version of Deming's quality process (Turicchi *et al.*, 2000) is often used for continuously improving safety. To optimize safety, the circular plan-do-check-act process can be used at all levels within the industry: from top (the aggregated multi-plant level) to bottom (the factory level). To be able to perform the Deming process at all levels, a safety-cycle structure can be established at each level and provided with communication and cooperation links between the different levels. These links are necessary for further optimization of the different levels of looping safety and for prevention of double elaboration of certain multi-leveled safety topics. Such a framework characterized by loop-level safety can be arranged as illustrated in Figure 4.2.

Plants follow the plan-do-check-act loop at installation, at factory and at plant level because of their acquired know-how of internationally accepted business standards such as the ISO 9000 series and the ISO 14 000 series, addressing quality and environmental management systems respectively. The OHSAS 18 000 series, the international Occupational Health and Safety management Assessment System specification that empowers an organization to control its occupational health and safety risks and improve its performance concerning those risks, is used to develop guidelines for working out safety-management systems at plant level. OHSAS is well known in the chemical industry and hence, some degree of basic standardization already exists in this specific industry. Thorough documented and well implemented safety systems at installation, at factory and plant level are readily available.

Regrettably, an international accepted standardization document addressing security management for industrial activities is not available.

Moreover, despite the increasing potential accidental as well as intentional danger represented by many industrial chemical sites worldwide where ever more hazardous chemicals are treated, multi-plant- or cluster-specific safety and/or security guidelines do not exist nor does an elaborated plan-do-check-act cycle at the multi-plant level.

4.2.2
Safety-Management Systems

Three different cooperation situations can occur in an industrial area. Hence, three types of safety-management-system development approaches can be distinguished, depending on the will of the plants to cooperate. First, if the will to cooperate does not exist between neighboring plants, each plant draws up its own individual plant SMS (PSMS). Secondly, if the plants are willing to cooperate fully and exchange all possible information necessary to minimize chemical risks in their area, a joint SMS (JSMS) can be drawn. These two situations can be regarded as the extreme cases of the cooperation continuum. However, the first option is by no means optimal from a social point of view since the companies do not take into consideration the fact that they belong to the same industrial area, thereby ignoring some very important low-frequency risks. The existence of major chemical risks as a result of other plants processing hazardous materials in their vicinity is not taken into account by the preventive and protective measures. This possibility is representative of the large majority of existing situations in the chemical industry today. The second option is to be preferred from a social point of view, although arguments can certainly be found against this way of operating. For example plants' activities can be extremely diverse and it may be sub-optimal to elaborate one JSMS for such diversity in activities. More people from different companies thinking about safety (management) also leads to more points of view, increased creativity, more innovative concepts and solutions, *etc.* Moreover, several arguments explain why this option might also be utopian: plants have often put a lot of time and effort into building their own plant SMS and therefore are often not willing to switch to a new joint SMS; some data must be regarded as highly confidential (even for cooperating companies); often a lot of distrust and competitiveness has existed (and still does exist) between the plants for a long time; *etc.* The third pos-

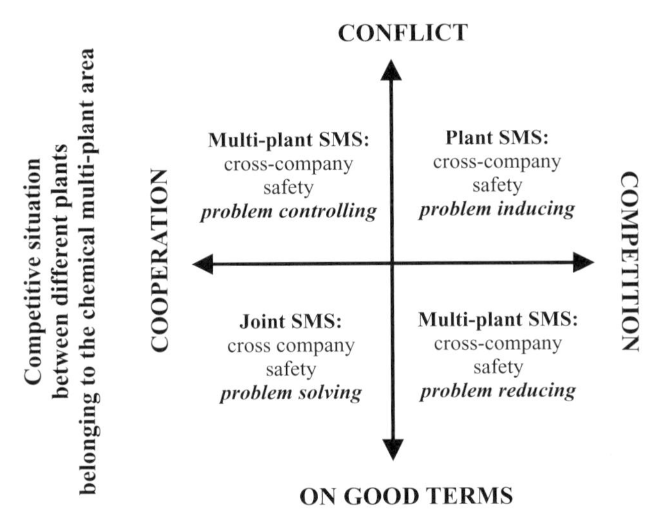

Figure 4.3 Categorizing cooperation situations within chemical clusters.

sibility is situated somewhere in the middle of the cooperation continuum. In this case, when the adjacent chemical plants are willing to cooperate, at least on a limited scale, a multi-plant SMS can be elaborated. To date, no such multi-plant-safety-management system exists in the chemical industry. For a better understanding of the different possible situations of neighboring chemical companies, a plant-cooperation spectrum can be drawn as in Figure 4.3.

This figure illustrates the spectrum of cooperation behavior of adjacent plants within chemical clusters and associates the type of safety-management system with it. Two competing plants having a social conflict situation might lead to a cross-plant safety problem-inducing situation whereby the companies are incited to restrict themselves strictly to the implementation of individual plant SMSs. Although most neighboring companies do not find themselves in such a conflict-competition situation, chemical plants today still have individual plant SMSs only as there is, as already mentioned, regrettably no other option available to the chemical industry nowadays. If the adjacent plants are on good terms and are also already cooperating, a joint SMS could be installed in this area. In fact, a joint SMS can be drawn in exact the same way as a plant SMS since the safety measures taken by the constituent plants can be regarded as those of one big plant and the features of the joint SMS are the same for all participating companies. In most cases, however, in practice the situation will be a relationship between the single plants that is characterized by the installation of a multi-plant-safety-management system. The latter gives directive to the participating companies as to how to minimize multi-plant chemical hazards, in this way optimizing continuous improvement of safety in the industrial area.

To make use of the know-how already available in single plants, the multi-plant SMS of a certain multi-plant initiative can be based on the individual plant-safety-management systems of those companies belonging to this multi-plant initiative. In the first stage, the various participating plants give input to the multi-plant SMS guideline. In this way, the multi-plant SMS adds divergent company processing activities and divergent views of the individual plant-safety managers to the multi-plant safety items list. Multi-plant – typical safety issues can be collected as well at this stage. The second step offers a multi-plant tailor-made directive for application within every individual plant. The existing PSMS is adapted if necessary and extra multi-plant safety issues are implemented in the plant. For a better understanding of the methodological concept, see Figure 4.4 that illustrates the situation for the chemical cluster represented in Figure 4.1.

A multi-plant-safety-management system consists of subjects raised in the plant-safety-management system and additional multi-plant-safety topics. By using this approach, the multi-plant-safety-management system is characterized by the same safety implementation features as every plant-safety-management system and furthermore uses these common features to address multi-plant safety issues as well.

4.2.3
Security-Management Programs

From a multi-plant perspective, managing safety and managing security in fixed chemical sites display some significant differences. Various guidelines are available for chemical plants to deal with different types of safety risks and to develop a PSMS. These guidelines can vary from country to country and from (organizational) culture to culture. By contrast, the concept of enhanced security (risk) management has only very recently been thoroughly dealt with by the chemical industry and thus there is only a small degree of diversification of security codes of good practice and security guidelines. In fact, companies only relatively recently became fully aware of the need to develop a solid security-management program and hence not much hands-on practice has yet been established. Moreover, theoretical studies on chemical-security optimization and chemical-security management have also only lately gained interest and importance. Therefore, tuning the existing plant-security-management programs into a multi-plant-security management program (M-PSMP) seems a somewhat more feasible task than shaping a multi-plant-safety-management system.

The next section describes how continuously improving safety and security topics at multi-plant level can be organized according to a tentative organisational initiative.

4.2.4
Setting Up a Multi-Plant Initiative

Depending on the severity of potential (accidental or intentional) accident consequences, plants' top management may decide to establish a multi-plant initiative

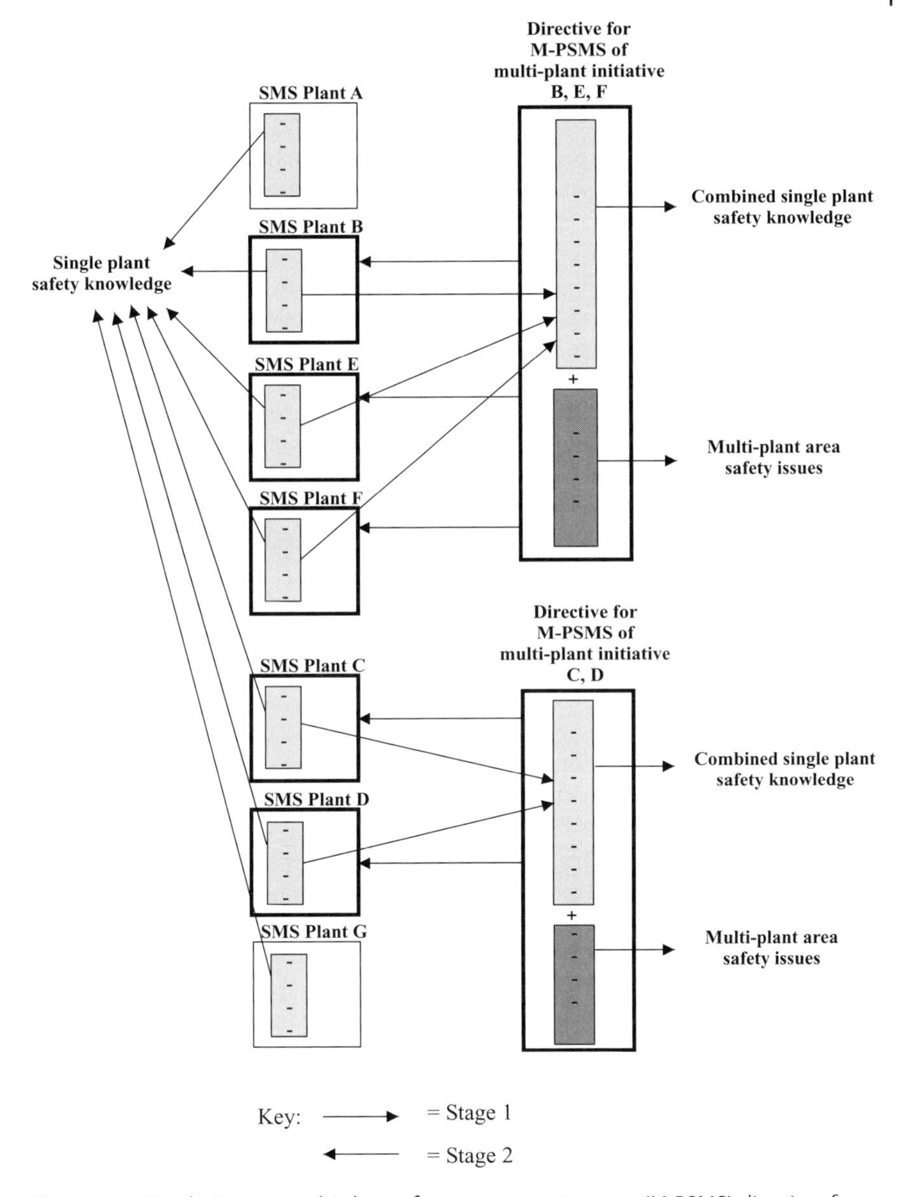

Figure 4.4 Developing two multi-plant-safety-management system (M-PSMS) directives for the case of two multi-plant initiatives within the seven-plant chemical cluster of Figure 4.1.

to continuously enhance safety and security of the plants belonging to this initiative.

The strategic goals as laid down in a safety policy are implemented through the use of a safety-management system. The latter comprises a written safety policy document indicating the plant (or several plants) management's commitment towards the underlying principles governing safety and health decisions. The plant (or several plants') security-management program fulfils an equivalent role regarding the corporate (or multi-corporate) security policy. These safety and security (S&S) policies are a reflection of the single-company S&S culture or the multi-plant S&S culture. In the multi-plant directive, multi-plant S&S requirements and S&S standards should be included with the objective of "translating" these to plant level and ultimately to working procedures at factory and at installation level.

Within a plant or at "joint plant" level, it is very important that these policies are clearly communicated to employees and relevant contractors in order to make them aware of their roles, responsibilities and accountabilities with respect to plant or "joint plant" safety and security. In the case of several plants, the safety and security policy documents are of a more holistic nature, emphasizing the important role of commitment, communication, cooperation, multi-plant responsibilities, domino or cross-company accident prevention, joint emergency preparedness, joint safety training, joint fence control, *etc.*

Such policies are best reviewed regularly and updated in the light of incident experience and any relevant changes in safety and/or security knowledge, technologies, laws and regulations.

The objectives for developing safety and security management guidelines, whether to enhance plant/joint S&S or multi-plant S&S, are first to minimize the likelihood of an accident, secondly to mitigate as conscientiously as possible potential consequences of accidents through emergency planning, land-use planning and risk communication, and thirdly to limit the eventually adverse consequences to health, the environment and property in the event of an accident. In the case of multi-plant-aimed management guidelines, the term "accident" denotes a major cross-company accident. To meet these difficult goals, the safety-management system and the security-management program address the appropriate actions that should be taken by a plant or by a group of plants, and by public authorities, communities and other parties involved. To pro-actively reduce future incidents, it also includes guidelines for implementing actions learned from the experiences of past accidents and other unexpected events.

Summarizing, a plant/joint or a multi-plant-safety-management system or security-management program is characterized by prevention, preparedness, response and follow-up of accidents caused by the equipment operated, or people operating within the premises of the plant or within the premises of several plants.

Responsibility for the issues to be added to the PSMSs and PSMPs for developing multi-plant S&S guidelines can be assigned to a semi-independent institute (being part of the organisational multi-plant initiative). In this way, none of the participating companies is favored and the reluctance of companies to give the required data is reduced. In fact, such an institute should be responsible for developing, maintaining and revising multi-plant S&S management systems as well as monitoring their implementation. To do this, relationships between various types of organizations have to be defined at different levels. Moreover, responsibilities and decision makers have to be assigned within these relationships. Hence, designing such extended S&S management systems is by no means an easy undertaking.

To organize a complex hybrid structure such as a part of a chemical cluster consisting of a group of organizations of the cluster (or in the ideal case the entire chemical cluster comprising all plants of the cluster), the term "organisational initiative" needs to be unambiguously defined. An organisational initiative is considered as a *system*. According to Daft (2001), a system (such as a chemical plant) can be defined as a set of interacting elements that acquires inputs from the environment, transforms these inputs, and discharges outputs to the external environment. The need for inputs and outputs reflects the dependency on the environment. Interacting elements imply that chemical plants belonging to an organisational initiative depend on one another and must work together, in this case on multi-plant safety and security issues. Moreover, a system is made up of several subsystems (in the case of chemical plants called the "departments"), performing the specific functions required for organizational survival, such as the production department, the maintenance department, human-resource department, logistics department, environment department, management, and safety and security department. The production department processes the inputs and produces the outputs of the organization. The maintenance department is responsible for maintaining the smooth operation of the production processes of the organization. The human-resource department deals with the organization's personnel issues, as well as with skills, training and education. The logistics department is involved in streamlining the existing supply chains of the organization. The environment department is concerned with all organizational topics (substances stored or processed, processes used, *etc.*) having an environmental impact both within the organization and external to the organization. Management (top and line) coordinates and directs the aforementioned departments of the organization. Management is also responsible for safe and secure operation within chemical plants and line management in the end makes up the balance between production goals and safety targets. Although safety and security support is part of the general management of an organization, it is at the same time independent from other parts of the organization. Safety and security support has some specific safety-related and security-related tasks, such as carrying out risk analyses, vulnerability assessments, drafting and updating safety and security documents, safety and security auditing, safety and security training, safety and security communication, emergency planning, *etc.* Hence, due to its independent nature, safety and security (managerial and operational) support is regarded as a distinct department, responsible for safety (inclusive health) and security issues in all the other departments, including management. Within an organization, the departments are interrelated, they cooperate and they exchange a variety of information to achieve the organization's goals.

Since every chemical organization consists of departments having approximately the same know-how, sharing information between such sub-organizational departments in a multi-plant context can be justified. To the mutual advantage of these departments, their expertise and experiences can be expanded to a multi-plant level. Figure 4.5 illustrates the organizational concept that can be used for establishing a multi-plant initiative.

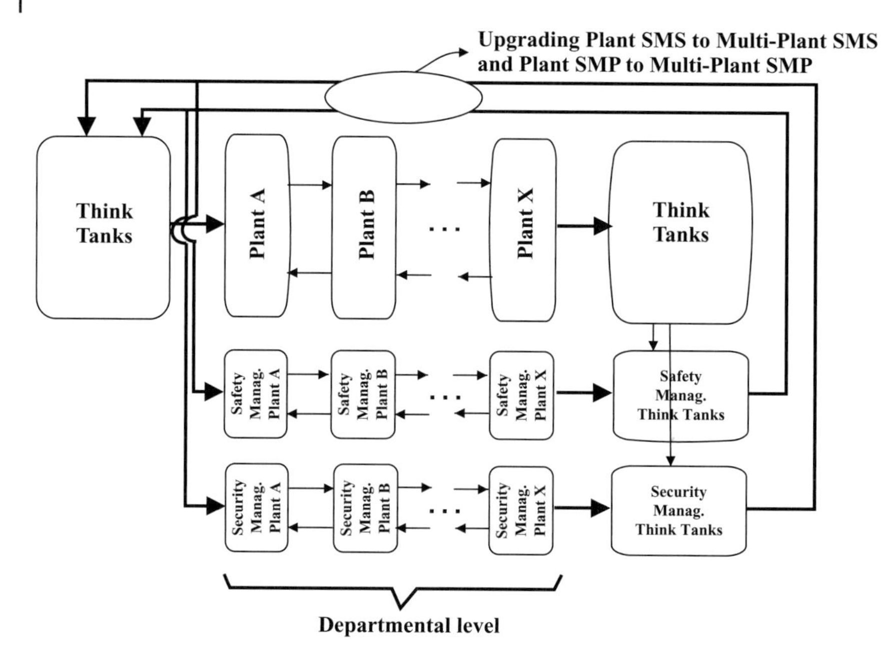

Figure 4.5 Setting up a multi-plant initiative.

In Figure 4.5, the multi-plant area is composed of a number of companies: *A, B, C, ..., X.* In the proposed conceptual framework, think tanks act as catalysts enhancing communication (gathering, structuring and providing information of all kind related to safety and/or security) in between individual plant departments and between these departments and cross-company safety/security management.

Four levels of analysis normally characterize organizations (Daft, 2001). The individual human being is the basic building block of each organization. The next higher system level is the department. These are collections of individuals who work together to perform group tasks. The next level of analysis is the organization itself. An organization is a collection of departments that combine to form the total organization. Organizations themselves can be clustered into the highest level of analysis, which is the inter-organizational or multi-plant set (all companies together within an industrial area are called the "cluster") and the community. This inter-organizational set is the group of organizations with which a single organization interacts. Figure 4.6 illustrates the four levels of analysis considering the case of the multi-plant initiative of three companies B, E, and F from the exemplary chemical cluster from Figure 4.1 composed of seven companies.

To describe an optimizing approach that deals with multi-plant safety and multi-plant security, a focus on the inter-organizational level of analysis is needed, paying special attention to organizations and departments. A semi-independent institute (as a part of the organizational initiative), hereafter called the "multi-plant council" (MPC), could streamline this multi-organizational level of analysis.

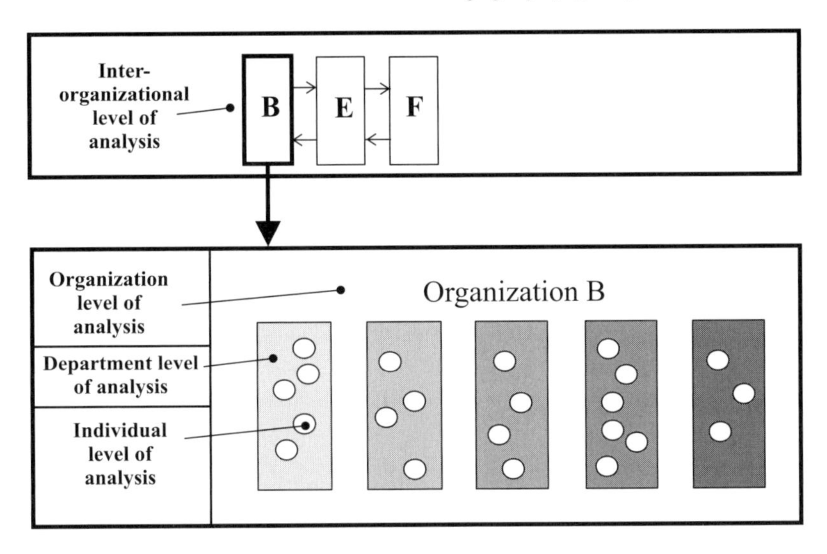

Figure 4.6 Four levels of system analysis.

Empirical research shows that chemical companies recognize the necessity for improved cooperation (Reniers *et al.*, 2005a). Companies are convinced of the safety-maximizing synergy effects of cross-company safety-risk analysis, but at the same time openly question the feasibility of more intensive cooperation for five main reasons. First, companies belonging to an international group with standard safety methods are often obliged to use these methods. Secondly, companies with divergent core activities need to be convinced of the safety gains of joint safety management. Several companies participating to the survey conducted by Reniers were, however, convinced that joint training courses and safety drills would improve safety. Thirdly, the desire to collaborate is often limited by practical problems, such as the procedure to purchase personal safety equipment. The fourth reason is the division of the costs of joint prevention measures, especially where mutual risks are not equally divided over the plants and are difficult to measure. The fifth main reason for hesitating to intensively collaborate with nearby plants may be summarized as liability questions. These considerations and the confidentiality of company safety data are the major hurdles facing collective safety-risk analyses in the chemical sector.

Cooperation between distrustful rival businesses can be enhanced if they are faced with a common enemy (e.g., terrorism). Security issues have the advantage of fulfilling the latter requirement. However, security management also works to prevent highly confidential company information from becoming known to competitors. Moreover, costs made to assure joint plant security also seem to be a major practical problem to overcome. Therefore, although chemical businesses face new security threats and respond to them with relatively new and fairly cross-enterprise standardized security vulnerability analyses, improved inter-firm security collaboration is subject to the same obstacles as those mentioned for enhancing safety collaboration.

Commitment and communication are key factors in enhancing cooperation and information exchange to improve internal and external safety and security. Once multi-plant S&S commitment and multi-plant S&S communication are accepted as key company notions, multi-plant safety and security cooperation can be streamlined as explained in the next section, giving recommendations to overcome the hurdles mentioned by Reniers *et al.* (2005a).

4.3
Plant-, Joint- and Multi-Plant-Safety and -Security-Management Stakeholders

4.3.1
Introduction

As indicated in Chapter 3, a plant-safety-management system (PSMS) aims to ensure the various safety risks posed by operating the facility are always below pre-defined and generally accepted company safety risk levels. A plant-security management program (PSMP) fulfils equivalent needs to control the company security risks. Effective management procedures adopt a systematic and pro-active (and thus anticipative) approach to the evaluation and management of the plant, its products, its supply chain, its maintenance schemes, its human resources, *etc.*

To enhance process safety, the PSMS considers safety features throughout process selection, process design, plant realization, commissioning, beneficial production and decommissioning. To enhance workforce safety, both personal and group safety equipment is provided, training programs are installed, and task capabilities are checked. Arrangements are made to guarantee that the means provided for safe operation of the industrial activity are properly designed, constructed, tested, operated, inspected and maintained and that persons working on the site (contractors included) are properly instructed.

To improve security, the PSMP identifies, analyzes and handles key security risk elements such as the *act* (e.g., economic, unlawful), *perpetrators* (e.g., individuals, groups, states), *objectives* (e.g., competitor advantage, political, personal gain), *intended outcomes and motivations* (e.g., financial losses, loss of lives, fear, frustrations), *targets* (e.g., individuals, installations, groups of individuals, plants, business assets), and *modus operandi* (e.g., spying, theft, bombing, hijacking).

Multi-plant-safety- and security-management procedures need to be constructed by (i) adding up the subjects as addressed by the plant SMSs or the plant SMPs from the companies forming part of the multi-plant initiative and (ii) discussing specific multi-plant safety and security topics.

Various guidelines and best practices for designing and implementing safety-management systems in individual chemical plants are available: for example, Code of Practice on Safety Management Systems for the Chemical Industry (OPITSC, 2001), Process Safety Management (Canadian Society for Chemical Engineering, 2002), Guiding Principles for Chemical Accident Prevention, Preparedness and Response (OECD, 2003), OECD General Guidance on Risk Management Programs for Chemical Accidents Prevention (OECD, 2004), Occupational Health and Safety Assessment System (OHSAS 18001:2007), *etc.* The requirements and recommendations of these codes of good practice that apply to individual major hazard plants in the process industries are compared, analyzed and validated in the light of multi-plant chemical safety.

Codes of good practice for designing and implementing security-management programs in individual chemical plants exist as well and are used in the discussion how to further enhance security in several plants belonging to a cluster. Some examples are Site Security Guidelines for the U.S. Chemical Industry (cf. Responsible Care® Security Code of Management Practices) (American Chemistry Council, 2001), Guidelines for Analyzing and Managing the Security Vulnerabilities of Fixed Chemical Sites (CCPS, 2003), General Security Risk Assessment Guidelines (ASIS, 2003), Protection against Terrorism (MI5, 2005), *etc.* These guidelines applying to individual chemical companies are compared, analyzed and validated in the light of chemical security on a multi-plant scale.

Four indispensable features for establishing a single- or multi-company SMS/SMP can be listed: (i) parties involved; (ii) safety and security policy–objectives; (iii) list of actions to be taken (practical recommendations); and (iv) implementation of the system.

In the next sections, the author's view on how to innovate this knowledge to construct multi-plant-safety and -security-management procedures is discussed and recommendations are formulated.

4.3.2
Parties Involved

The guidelines for plant-, joint- and multi-plant S&S management systems should include all stakeholders involved in safety and security management in hazardous industries, each of them having different responsibilities. Therefore, the drawing up of a PSMS, a JSMS or a M-PSMS takes into account the viewpoints, the tasks and the responsibilities of several groups, as represented in Table 4.1.

A great deal of attention is paid to the participation and the involvement of all plant's, respectively plants', employees in drawing up and implementing a PSMS or a PSMP, respectively a JSMS or a JSMP, as they have extensive hands-on knowledge that can be used in the process. In addition to the plant labor force and plant line management, top-level management and the multi-plant (safety & security) council (MPC) in particular are very important stakeholders when it comes to implementing a M-PSMS and a M-PSMP. At present, informal gatherings of companies belonging to the same industrial cluster are (slowly) evolving towards more formal organizations discussing operational practices. Examples are the "Deltalinqs University" that was founded in the port of Rotterdam in 2001 and the "Delta Process Academy" (DPA) founded in the port of Antwerp in 2005, ports hosting two major industrial chemical clusters in Europe. These initiatives express the willingness of the major players to cooperate more intensively and on a larger scale in the field of operational safety.

Table 4.1 Stakeholders in safety-management systems.

Stakeholder category	**Plant**		**Public**	**Public authorities**	**Emergency responders**	**Other**	**Neighboring plants**	**Multi-plant council**
Stakeholder group examples	Owners, shareholders	Managers, labor force	All the people who could be influenced by a hazardous event (external to the plant or cluster of plants)	Authorities at national, regional and local level	Fire fighters and medical emergency responders	Research institutes, labor organizations, and business organizations	Plant safety and plant-security managers from companies belonging to the same industrial area	Consultancy agencies, independent body
PSMS	√	√	√	√	√	√		
JSMS	√	√	√	√	√	√	√	
M-PSMS		√	√	√	√	√	√	√

To indicate the novelty of the approach explained in this book, a conceptual comparison was carried out with three existing approaches of intensive operational collaboration within chemical areas.

The first approach discussed is the ValuePark® concept developed by The Dow Chemical Company (ES2, ES3, 2008). Dow implemented the concept for example in Terneuzen (The Netherlands) and near Leipzig (Germany). These projects by Dow are inspired by the "major user" principle, indicating the industrial area to be dominated by a so-called major user (i.e. The Dow Chemical Company), and to serve as an investment area for medium-sized enterprises. The approach of the concept is to get preferred investors jointly settle in the ValuePark with the objective of creating synergy effects and improving the competitive situation on the market. Three main synergy effects are envisioned: (i) integration of material flows and logistic services, (ii) joint use of available infrastructure, and (iii) decreasing the necessary investment capital. Concerning safety and security within the ValuePark®, a distinction is made between mandatory services (fire department and security services) and optional services (medical services, industrial safety, industrial hygiene, and staff training) provided by the major user. Hence, the concept is strictly concentrated on operational collaboration as regards linked production and linked delivery of services (e.g., safety services, security services) within an industrial area provided by the so-called major user of the area. Therefore, the objectives of the ValuePark® cooperation projects regarding safety do not aim to continuously improving the area's safety and security through integrated and equally appreciated operational and procedural collaboration of chemical company neighbors, but the "smaller firms" depend on the "major user".

The second exemplary approach concerns an industrial symbiosis at Kalundborg, Denmark. The philosophy behind the symbiosis of Kalundborg is that companies exploit each other's residual or by-products on a commercial basis. Hence, the operational cooperation results in reduced consumption of resources and a significant reduction in environmental strain. Although the industrial symbiosis implies co-existence between diverse organizations in which each may benefit from the other, the individual agreement within the collaboration concept is based on commercial (and to a lesser extend environmental) principles. Some conditions for cooperation are believed to be required for the Kalundborg experience (ES4, 2008), that is (i) the companies must fit each other, (ii) the companies must be located near each other, and (iii) there must be openness between the companies. As a result, this industrial symbiosis initiative does not focus on cluster process safety issues at all and cannot be generalized to other industrial complexes (because there should be openness and mutual trust between the organizations). Therefore, the Kalundborg project cannot be implemented in "classic" cases such as, for example, the Houston, the Antwerp or the Rotterdam chemical clusters since in these major clusters there is much less openness, mutual trust, *etc.*, than in Kalundborg.

A third approach already used in the industry, is the Chemelot initiative in Geleen, The Netherlands. This initiative is a unique chemical and materials community that thrives on the concept of open exchange of ideas between companies. Chemelot has emerged around one central concept, that is, to bring together the knowledge and skills normally found only in major organizations, and to apply these within a flexible community of small and large chemical businesses. According to the Chemelot community, their initiative leads to continuous innovation in technology, products, applications, production processes, business models, organization forms, *etc.*, in cooperation with, and with the help of, other companies. Obligatory services include fire-fighter services and security services. However, the initiative does not employ an organizational model for continuously improving safety and security within the area. Moreover, although very interesting for SMEs (regarding the open innovation concept), the initiative does not seem to gain much trust and openness of many large chemical multinationals. The Chemelot concept cannot be applied to existing clusters such as Houston, Antwerp, Rotterdam, *etc.*, due to the extreme diversity in company cultures in such major clusters.

In this book, we discuss a concept for continuously improving safety and security within an industrial area, irrespective of the relationship between the participating companies of the cluster. The added value and the validity of the approach elaborated thus comprise the focus on multi-plant safety and security collaboration between equally appreciated partners, and the general applicability of the suggested concepts, independent of the participating companies' existing cultures.

More profound safety and security cooperation requires intensive collaboration at operational, tactical and strategic level (supported by top-level management) where it should be guaranteed that specific data is treated confidential. These objectives can be achieved by a multi-plant council composed of company representatives and independent experts (see also Section 4.3.3). The concept of splitting the multi-plant council into two parts was cross-checked by S&S experts belonging to public authorities as well as S&S experts from private companies. All experts agree concerning the soundness and the potential of the suggested approach. Using existing best industrial practices from single chemical companies and integrating them into a best cross-industrial practice for continuously improving safety and security within industrial areas is believed by the experts to be the only possible methodology/way for bringing theory into practice. The approach of the multi-plant council was considered to be necessary to guarantee the confidentiality of information to be respected, thereby lowering the suspicion of participating plants and making the concepts of a M-PSMS and a M-PSMP more feasible. The suggested approach indeed might lead to a safety and security situation of an industrial area of the next generation.

4.3.3
The Multi-Plant (Safety & Security) Council (MPC)

To structure safety and security issues at a multi-plant level, multi-plant-related coordination is organized by a "multi-plant council", grouping organization representatives from participating plants and independent delegates. Hence, the multi-plant council includes two main parts as illustrated in Figure 4.7.

The first main part consists of two sub-parts and is composed of plant safety representatives (sub-part 1) and plant-security representatives (sub-part 2). It has

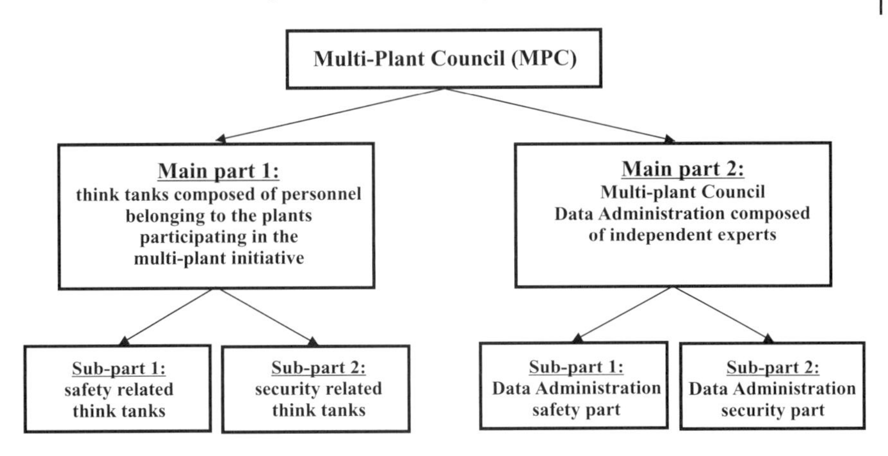

Figure 4.7 Organogram of the multi-plant council.

a typical counseling function, formulating safety and security recommendations as a result of joint think-tank brainstorming and communication sessions. The other part, the multi-plant council data administration, consists of two sub-parts as well: one for safety and one for security. The MPC data administration safety part is composed of independent consultants (i.e. impartial knowledgeable personnel) responsible for administering all necessary (confidential) safety-related information gathered from the different plants of the cluster. Although the MPC data administration security part has a structure equivalent to the safety sub-part, both administrations are separated and work independently from each other due to the extremely high confidentiality of plant-security data.

An MPC think-tank organizes brainstorming sessions for the plant departments of production, maintenance, human recourses, logistics, environment, management, safety management and security management. The number of participants of such a think-tank is deliberately limited with a view to maximizing output efficiency. If the number of participating plants in the multi-plant initiative is too large, a method of systematic alternation of representatives can be used. As a guideline, a maximum of eight representatives per think tank are proposed: seven department representatives (after a fixed period of time one representative after another alternating with department representatives from the other companies) and one independent consultant with expertise in the department field. The MPC safety-management think tank and the MPC security-management think tank aim at achieving integrated preventive multi-plant safety and security, respectively, by drawing and proposing standardized procedures (together called the M-PSMS and the M-PSMP, respectively) based on plant department recommendations and added multi-plant safety issues and multi-plant security topics. The crucial role of the safety- and security-management think tanks is reflected in their composition, that is permanent multi-plant safety and security specialists from the most dangerous companies are added to the small group of eight. The safety- and security-management think tanks maximum consist of 12 group members, the numbers again being limited for the same reason as given previously. The M-PSMS/M-PSMP is translated and implemented at individual plant level by the departments. Hence, a continuous improvement of drafting the different parts of the M-PSMS/M-PSMP and the PSMSs/PSMPs is achieved by optimized communication and cooperation of every department of the plants participating to the multi-plant initiative.

The collection of multi-plant data (e.g., incidents, accidents, *etc.*), for example, employed for inspecting plants compliance with multi-plant recommendations, for auditing plants towards multi-plant safety and security requirements and for performing all kinds of administrative tasks, can be supported by the MPC data administration. Such an administration works closely together with the safety management think tank and the security-management think tank and includes independent external safety, security, decision-support systems and inspection and auditing experts.

By splitting the multi-plant council into a part composed of plant personnel, that is, the think tanks, and a part composed of independent experts, that is, the data administration, a balance between confidentiality and information sharing is targeted. The think tanks and the multi-plant organizational platform ensure the continuous improvement in taking preventive measures as regards multi-plant topics and, to a lesser extent, single-plant safety and security topics. The data administration collects the necessary (confidential) technical installations (and processes) data, incident and near-incident data and other safety and security information and uses this information, for example, as input for executing decision-support systems (i.e. computer-automated software), audits, inspections, *etc.* Based on the output, the MPC safety data administration gives guidance and recommendations to the MPC safety-management think tank, as does the MPC Security data administration to the MPC security-management think tank. If calamities should occur having a possible impact beyond the originating company, the necessary data of all the plants is centralized in an MPC data administration databank and can be used without any delay. Based on Figure 4.5, a more practical scheme for optimized multi-plant organization is given in Figure 4.8.

Figure 4.8 illustrates the platform that can be used by the participating plants belonging to the multi-plant initiative of companies B, E and F proposed in Figure 4.1 to streamline their multi-plant organization and to enhance their multi-plant safety and security management, and taking into account the scheme from Figure 4.4 and Figure 4.5. In Figure 4.8, each of the three companies is composed of 8 plant departments (those departments having an impact on safety and security results are involved in the organisational platform). Plant departments, for example, communicate through e-mail and a multi-plant intranet. Think-tank sessions are periodically organized. Although the scheme of Figure 4.8 provides an extensive illustration of organizing the MPC plant representatives in an optimizing way, it fails to include the MPC data administration parts. To this end, Figure 4.9 is given.

Figure 4.9, combined with Figure 4.10a,b, illustrate the safety-management systems and security-management programs communication and cooperation

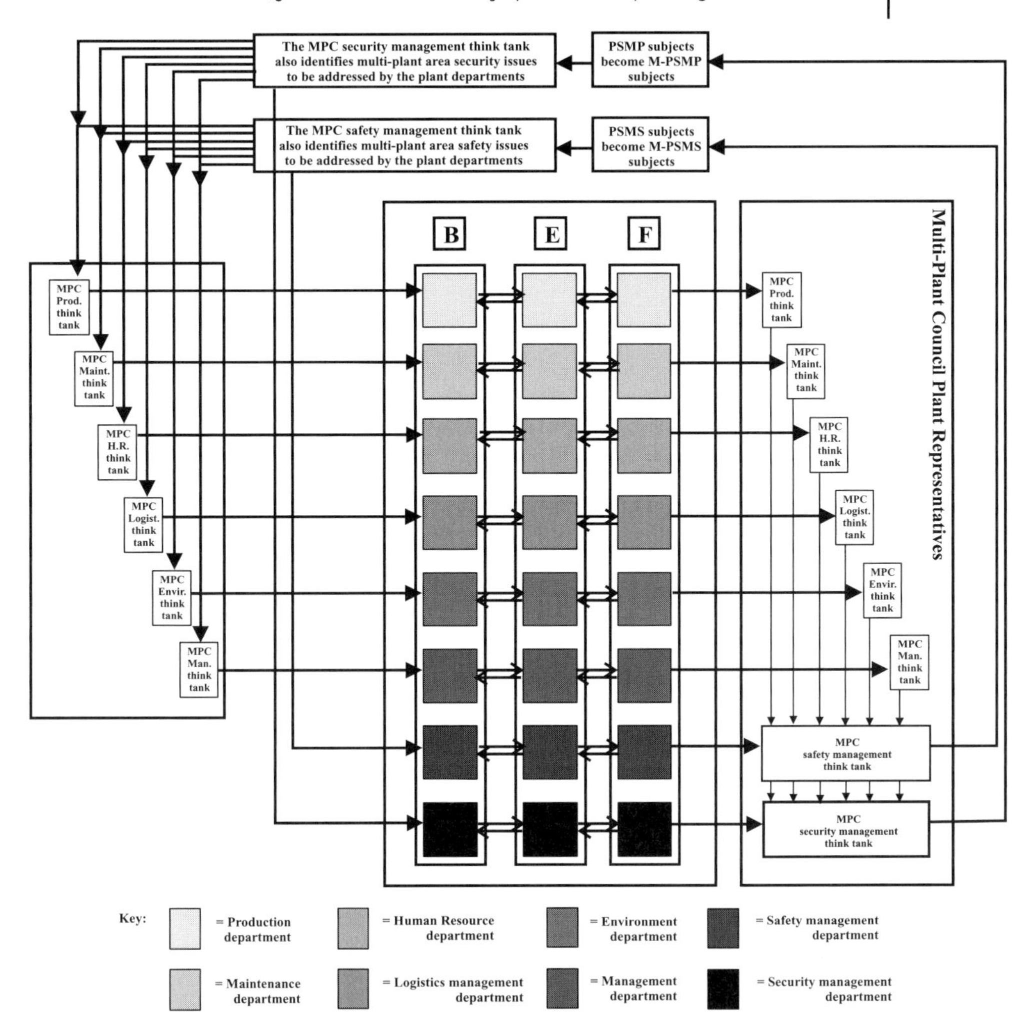

Figure 4.8 Platform for optimized multi-plant organization in case of multi-plant initiative composed of three companies *B*, *E*, and *F*.

procedure, offering a loop for continuous improvement of multi-plant safety and multi-plant security.

Figure 4.10a,b, suggest a non-exhaustive division of responsibilities of the different parts constituting the multi-plant council.

The multi-plant organizational platform and the multi-plant council as a part thereof guarantee the performance of processes entailing planning, establishment of goals and objectives, monitoring of progress and performance, analysis of trends and development and implementation of corrective actions to be continuously improved, both for safety and security, and at both plant and at multi-plant level.

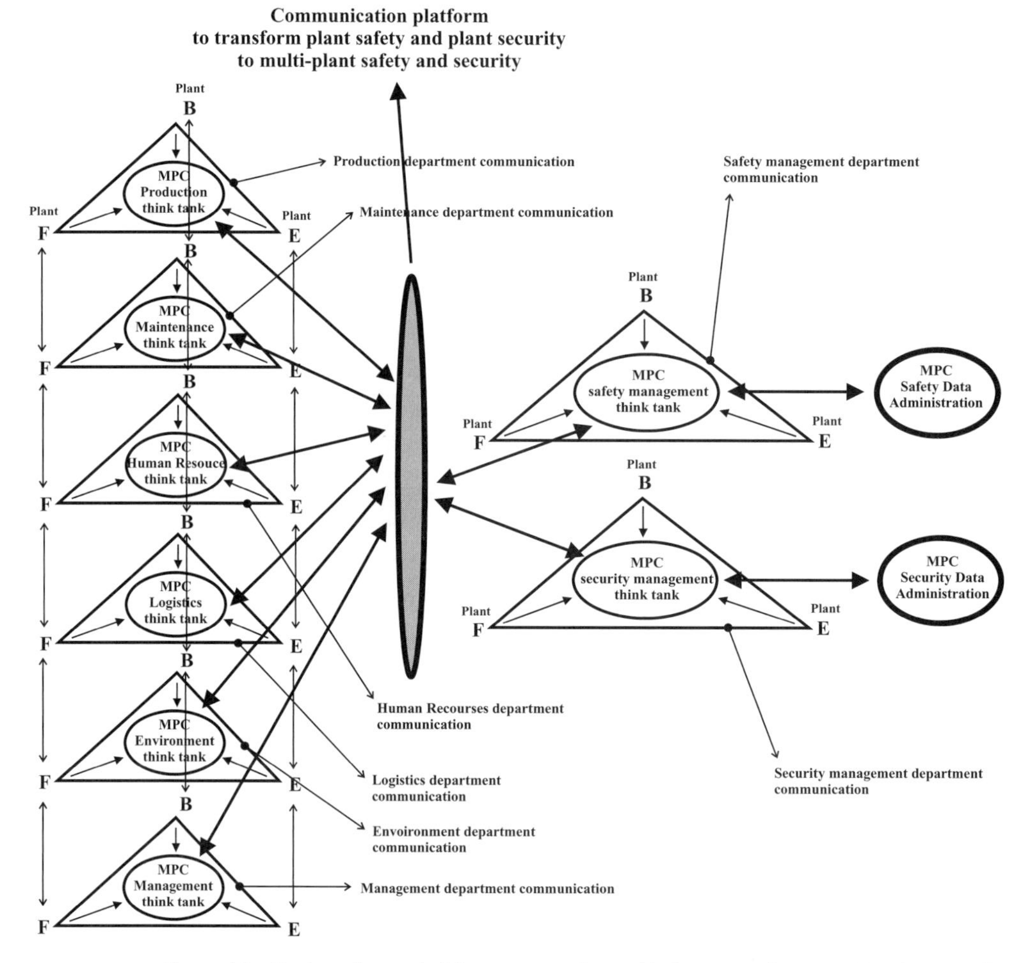

Figure 4.9 Design of an optimizing structure for multi-plant council composition in case of a multi-plant initiative composed of three plants *B*, *E*, *F*.

Figure 4.10 (a) Circle of continuous improvement of multi-plant safety; (b) Circle of continuous improvement of multi-plant security.

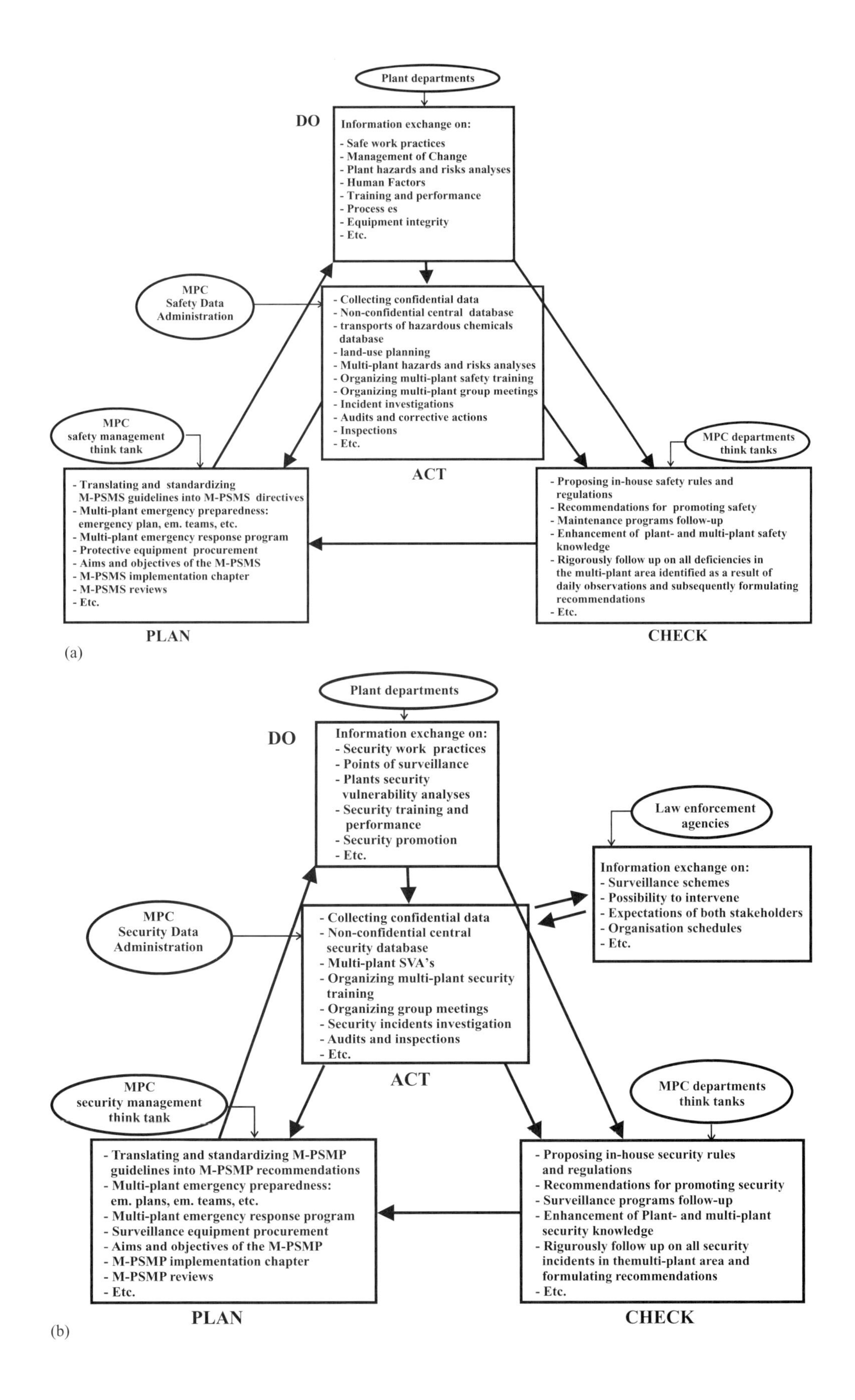
Plant departments
DO
Information exchange on:
- Safe work practices
- Management of Change
- Plant hazards and risks analyses
- Human Factors
- Training and performance
- Process es
- Equipment integrity
- Etc.
MPC
Safety Data
Administration
- Collecting confidential data
- Non-confidential central database
- transports of hazardous chemicals database
- land-use planning
- Multi-plant hazards and risks analyses
- Organizing multi-plant safety training
- Organizing multi-plant group meetings
- Incident investigations
- Audits and corrective actions
- Inspections
- Etc.
ACT
MPC
safety management
think tank
MPC departments
think tanks
- Translating and standardizing M-PSMS guidelines into M-PSMS directives
- Multi-plant emergency preparedness: emergency plan, em. teams, etc.
- Multi-plant emergency response program
- Protective equipment procurement
- Aims and objectives of the M-PSMS
- M-PSMS implementation chapter
- M-PSMS reviews
- Etc.
PLAN
- Proposing in-house safety rules and regulations
- Recommendations for promoting safety
- Maintenance programs follow-up
- Enhancement of plant- and multi-plant safety knowledge
- Rigorously follow up on all deficiencies in the multi-plant area identified as a result of daily observations and subsequently formulating recommendations
- Etc.
CHECK
(a)
Plant departments
DO
Information exchange on:
- Security work practices
- Points of surveillance
- Plants security vulnerability analyses
- Security training and performance
- Security promotion
- Etc.
Law enforcement
agencies
Information exchange on:
- Surveillance schemes
- Possibility to intervene
- Expectations of both stakeholders
- Organisation schedules
- Etc.
MPC
Security Data
Administration
- Collecting confidential data
- Non-confidential central security database
- Multi-plant SVA's
- Organizing multi-plant security training
- Organizing group meetings
- Security incidents investigation
- Audits and inspections
- Etc.
ACT
MPC
security management
think tank
MPC departments
think tanks
- Translating and standardizing M-PSMP guidelines into M-PSMP recommendations
- Multi-plant emergency preparedness: em. plans, em. teams, etc.
- Multi-plant emergency response program
- Surveillance equipment procurement
- Aims and objectives of the M-PSMP
- M-PSMP implementation chapter
- M-PSMP reviews
- Etc.
PLAN
- Proposing in-house security rules and regulations
- Recommendations for promoting security
- Surveillance programs follow-up
- Enhancement of Plant- and multi-plant security knowledge
- Rigurously follow up on all security incidents in themulti-plant area and formulating recommendations
- Etc.
CHECK
(b)

4.4
Practical Recommendations for Achieving Plant or Multi-Plant-*Safety* Loop of Continuous Improvement

4.4.1
Introduction

To create a solid safety-management system, two separate safety cycles should be elaborated: one aimed at preventing chemical accidents of any kind and one aimed at mitigating the consequences of a chemical accident in the unfortunate scenario of error. The necessary actions resulting from drawing up a safety management plan can thus be divided into three main categories: (i) prevention of (accidental) chemical accidents, (ii) mitigation of chemical accidents, and (iii) follow up and corrective actions. Therefore, these issues must be included in the PSMS/JSMS or the M-PSMS. In the suggested platform, these three categories are treated by the same think tanks. In this way, information concerning prevention, mitigation and follow-up of safety- and security-related topics within the multi-plant context is processed in an integrated manner.

4.4.2
Prevention of (Accidental) Chemical Accidents

The essence of accident prevention practices consists of safety data, hazard reviews, operating procedures and training. These elements need to be integrated into a safety-management document that is implemented in the plant or in the cluster on an on-going basis. To enhance implementation efficiency, the latter topics are divided into 11 subjects. The subjects recommended for drafting a PSMS or a JSMS, briefly described in the summarizing Table 4.2 at the end of this section, can be found extensively elaborated in available guidelines. For a better understanding of the additional feature recommendations required to elaborate a multi-plant-safety-management system, these topics are more thoroughly described hereunder.

4.4.2.1 Safe Work Practices

Gradual agreement over work practices, used in the different plants, is established. By achieving this, work practices are standardized step by step wherever possible and their effectiveness is continually optimized through evaluation of the past experiences in the participating plants of the initiative.

4.4.2.2 Safety Training

The necessity to periodically organize multi-plant training sessions emerges from the continuously changing environment of plants belonging to the multi-plant area and of the area's vicinity. Employees working in a hazardous environment are trained together with their colleagues in nearby plants. Such group safety training sessions should lead to a more efficient handling of for example, an

Table 4.2 List of actions for the prevention of chemical accidents.[3)]

Safety Items	PSMS/JSMS Short content:	M-PSMS Additional features:
1. Safe work practices	A description is given of (i) The tasks of every process operator; (ii) Safe process operation parameters which must be maintained at all time; (iii) Safety precautions.	Enhance cross-company knowledge about the work practices used in the different plants to improve work practice effectiveness through learning and to decrease domino risks through external information on potentially dangerous or vulnerable installations.
2. Safety training	Safety training is needed to create (i) Safety consciousness and commitment among new employees; (ii) Necessary long-term comprehensive on-the-job training or follow-up training.	(i) Create multi-plant safety consciousness by being aware of an accident escalation leading to a disaster; (ii) Multi-plant safety training sessions to prevent or to mitigate knock-on effects by acting "correctly" in the event that a "dangerous" incident/ accident has occurred in a nearby plant.
3. Group meetings	Stimulating communication and cooperation between plant management, employees and contractors about plant safety and health topics.	Stimulating communication and cooperation in cross-company meetings between plant representatives about multi-plant safety topics.
4. In-house safety rules and regulations	Covering the work operations or processes in the plant (clearly documented and communicated to all relevant employees and contractors).	In-house safety regulations have to take into account the possibility of a domino accident caused by an adjacent company.
5. Safety promotion	(i) To maintain awareness among employees and contractors of the importance of safety and health in the plant; (ii) To create a plant-safety culture.	(i) To increase awareness among employees of the potential danger of their plant as a result of its role as part of a chemical cluster, and thus of the importance of multi-plant safety; (ii) Creating a multi-plant-safety culture.
6. Contractor and employee evaluation, selection and control	To select, control and evaluate contractors working with hazardous installations.	A multi-plant contractor program establishing common basic safety performance indicators needs to be drafted.

3) These items are non-systematically listed and hence no particular attention should be paid to their sequential order. The item's sequence is merely a result of the available safety-management system guidelines.

Table 4.2 *Continued*

Safety Items	PSMS/JSMS	M-PSMS
	Short content:	**Additional features:**
7. Safety inspection, monitoring and auditing	In-house control of personnel compliance with regulatory requirements, in-house safety rules and safe work practices. The inspection program includes committee inspection, routine safety inspection, plant and equipment inspection and surprise inspection.	(i) Monitoring by the multi-plant council to control compliance with multi-plant regulatory requirements concerning multi-plant safety; (ii) Monitoring by the multi-plant council to control the inspection programs of the initiative's participating plants.
8. Maintenance regimes	To prevent and predict problems induced by a lack of maintenance. A maintenance program includes inspection programs of machinery, corrosion control programs, work practices and maintenance procedures, training, design specifications, long-term maintenance plan for periodic maintenance of critical equipment, and a control system for maintenance of critical safety devices.	Monitoring by the MPC to verify the effectiveness and the long-term solid character of the maintenance programs of the initiative's participating plants.
9. Hazard analysis	To identify all possible process and operational hazards for determining the mechanisms by which they could give rise to undesired events, and to evaluate the consequences of these events on health (including public health, the environment and property).	The multi-plant council has to (i) Identify cross-company consequence scenarios; (ii) Identify in a standardized way individual plant risk contours and identify multi-plant risk contours by the objective adding up of individual risk contours; (iii) Identify and analyze cross-company risks for prevention optimization.
10. Control of movement and use of hazardous chemicals	(i) Control program including procedures for proper use, handling and movement of hazardous chemicals; (ii) Through the use of material safety data sheets (MSDS) employees know how to correctly receipt, issue, distribute, handle and safely use the hazardous chemicals.	Communication (database) program for cross-company information exchange on the use, handling and movement of hazardous chemicals within plants forming part of the multi-plant initiative.
11. Documentation control and records/hazard communication	(i) To easily find safety procedures for every operation, machinery and equipment in the plant; (ii) To document accidents, near-misses and incidents and to communicate the lessons learned.	Drafting a multi-plant database with relevant safety and danger information (at plant level) related to escalation hazards. Such a database should be accessible to every plant forming part of the multi-plant initiative.

on-going domino accident due to a collective know-how regarding how to act correctly. Coordinating the organization of these training sessions belongs to the responsibilities of the multi-plant council.

Depending on the circumstances and on the subjects, employees from different plants are trained either separately or jointly. The multi-plant training programs are reviewed periodically by the multi-plant council. In view of this, each company reports changes made to operational procedures.

4.4.2.3 Group Meetings

The multi-plant council is responsible for organizing group meetings. To achieve a smooth procedure at meetings, it is recommended that every plant organizes an internal group meeting about multi-plant safety issues to which the plant employees concerned are invited. Afterwards, a plant representative is delegated to the multi-plant group meeting to report the plant's own particular point of view.

4.4.2.4 Pursuing In-House Safety Rules and Complying with Regulations

In-house safety rules take into account the possibility of an accident caused by a neighboring company. Furthermore, every company potentially able to cause a multi-plant accident should exchange information with external companies concerning such a risk. The exchanged data would have to be used to create an emergency plan, to formulate a safety report and to establish a safety policy concerning the prevention and controlling of multi-plant accidents. An internal emergency plan and a contingency plan also should be drafted regarding external domino risks.

4.4.2.5 Safety Promotion

Safety training and plant group meetings are perfect tools to increase employees' awareness of the danger of hazardous installations and domino risks. However, to create a continuous awareness of the importance of multi-plant safety, other tools such as signs and posters will indicate the cross-company danger of some installations and substances.

4.4.2.6 Contractor and Employee Evaluation, Selection and Control

The participating companies establish an equivalent contractor assessment system based on the same basic safety-performance indicators. In this way, personnel under external management, but working within different plants participating in the multi-plant safety enhancing initiative, are treated, evaluated and rewarded in the same manner (concerning safety issues) as internally managed personnel.

4.4.2.7 Safety Inspection, Monitoring and Auditing

The multi-plant council is responsible for multi-plant safety inspection and multi-plant safety auditing. First, the council monitors plant compliance with multi-plant safety regulations (e.g., concerning external domino-accident prevention). Secondly, the council inspects the effectiveness of the individual plants' safety inspection programs (described in the plant's safety-management system).

4.4.2.8 Maintenance Regimes

A plant maintenance program plays a crucial role in the prevention of multi-plant accidents. Therefore, such plant programs are reviewed regularly by the multi-plant council and experience-based recommendations are made by the council.

4.4.2.9 Hazard Analysis

To perform a cross-company hazards analysis, it is vital that the multi-plant council has at its disposal all relevant information from the different participating companies. Information such as hazardous chemicals material safety data sheets (MSDS), the location of these chemicals, *etc.* has to be given to the data administration part of the multi-plant council. Based on these confidential data, a multi-plant domino hazard investigation can be carried out by an independent and certified organization and/or by personnel from the plants involved (see Chapter 5 for the suggested frameworks in this book). The MPC also informs other stakeholders, such as the public and the public authorities, about the consequences of the identified scenarios on safety, health (including public health) and the environment. Analysis results are discussed in group meetings with safety managers from the different companies involved. If necessary, these managers will then transfer the information to their own employees to be sure that it is clearly communicated to all parties involved.

4.4.2.10 Control of Movement and Use of Hazardous Chemicals

To ensure that all information on the use, handling and movement of different kinds of hazardous chemicals in the initiative's participating plants reaches the envisioned companies, a continuously adjusted database with non-confidential information is accessible to all the plants involved. The database is located on an information network shared by the different participating plants so that every company can carry out hazardous chemicals transportation and storage changes, while safety managers of the other plants belonging to the cluster are informed in real time.

4.4.2.11 Documentation Control and Records

In order to ensure that all companies have the same major and multi-plant accident prevention know-how, a central database managed by the MPC for gathering non-confidential as well as confidential safety information on an individual plant level about potential (theoretical and experienced) incidents possibly affecting other plants of the cluster can prove very useful.

Table 4.2 briefly discusses these eleven topics, comparing the existing recommendations for a PSMS/JSMS with the needs for a M-PSMS.

4.4.3 Mitigation of Chemical Accidents

In a PSMS/JSMS, for every factory of the plant/plants the potential emergency situations and their impacts are identified. Afterwards, a response plan is developed at factory and at plant/plants level with unambiguous roles and responsibilities for everyone. Furthermore, an emergency team is composed and procedures for raising alarms, initial response to emergencies, evacuation and rescue, capability of in-house resources, first-aid planning, *etc.*, are drafted. The ultimate plant-safety management system or joint safety-management system response plan will be tested through emergency exercises and simulations in the plant/plants.

In the case of an M-PSMS, possible escalation accident scenarios and their consequences are identified. Using these data, the participating companies elaborate a joint emergency plan. Such a plan includes listing the actions necessary in the event of a particular emergency situation. In the case of multi-plant safety issues, at least two companies per scenario are taken into account. Therefore, coordinating multi-plant emergency preparedness is a rather complex matter. The plant emergency plan described in each PSMS of the initiative's participating plants considers multi-plant emergency issues and the plans should be attuned to one another. Furthermore, the involvement of the fire brigade, the police and other authorized authorities is necessary.

The multi-plant emergency plan contains at the least automated early-warning systems (which play a key role in countering possible domino incidents) and multi-plant emergency centers that can act efficiently in an unexpected situation. Emergency teams have to be formed for the multi-plant area, eventually logically sub-divided into several smaller teams of sub-clusters encompassing the area of the multi-plant initiative. Dividing the multi-plant area into several sub-clusters (or "domino islands") can actually be an important security measure as well. Should new equipment be added to the main cluster, land-use planning guidelines are implemented and the MPC is fully responsible for all coordination activities.

An individual plant emergency-response program consists of four elements: an emergency-response plan, emergency-response equipment procedures, employee training and procedures to ensure the plan is up-to-date. The latter two items have already been discussed in this section. A plant emergency-response plan includes information about how to communicate with the public and the local agencies in the event of an unwanted incident, advice on first-aid and emergency medical treatment, and information about the plant procedures and measures for emergency response. Emergency equipment includes among other things the protective equipment for medical emergency response personnel. Procedures are written down about how and when to use this equipment, how to carry out maintenance and how to conduct regular inspections.

In a M-PSMS a multi-plant emergency-response plan, multi-plant emergency response equipment procedures, multi-plant emergency training and procedures in general are elaborated. They are drafted for several sub-clusters (belonging to the multi-plant constellation) to ensure that, should a domino accident happen in one of the sub-clusters (for example plants *E* and *F* of the multi-plant initiative comprising plants *B*, *E* and *F* (see Figure 4.1)) with limited or no consequences to other sub-clusters (for example plant *B* of the multi-plant initiative comprising *B*, *E* and *F*), the latter sub-clusters know their involvement in the emergency program, but remain the least affected in economic terms.

Therefore, distributing, handling and fixing responsibilities of necessary measurement, regulation and supply of equipment in the case of an emergency at multi-plant level is strictly regulated in the multi-plant emergency-response program.

Setting up an "emergency responder tool" can also be recommended. Such a tool would be able to collect and archive multi-plant data (e.g., on floor plans, utility shut-offs, geo-tagged aerial photographs, critical infrastructure characteristics, and other multi-plant area information) and disseminate that information via a web-based environment. Such a tool would represent a multi-plant mechanism by which fire departments of neighboring plants or of nearby communities, police departments, emergency responders, *etc.*, may obtain critical information in the event of a large-scale incident.

4.4.4
Follow-Up of Incidents, Incident Investigation and Corrective Actions

Past incidents can be a source of valuable information about site hazards and the steps that one needs to take to prevent accidents from happening. Often, the immediate cause of an accident is a series of other problems that need to be addressed to prevent a repetition of the incident. Therefore, in a PSMS or JSMS, procedures are set up guaranteeing that every incident will be reported promptly. These procedures will indicate the criteria for the type and degree of seriousness for incidents requiring more detailed investigation.

In order to establish a multi-plant incident database, all incidents taking place in participating plants are reported to the multi-plant council. By investigating and discussing these incidents in the light of multi-plant safety, the knowledge gained can be shared across the different organizations and will contribute to better overall accident prevention. Since different plants are involved, it is advisable to first organize an internal investigation at plant level with line managers, supervisors, safety personnel and safety committee members, thereby obtaining all relevant information. Afterwards, incident group meetings can be organized by the MPC with a view to reporting and to discussing cross-company-relevant incidents. In the follow-up of such potential inter-plant incidents, responsibilities of the individual plants are discussed by the multi-plant council. The latter, having merely a counseling function, offers pro-active advice to the participating plants.

Accidents and incidents that occurred five years previous to the formulation of the PSMS/JSMS are reported. Relevant data recorded for each incident are: date, release duration, type chemicals involved, quantity released, release source, on-site impacts, known off-site impacts, weather conditions (this can aggravate or mitigate the consequences) and the initiating events of the release. The objective of setting up this database is to determine the facts, conditions, circumstances and causes of chemical accidents on a plant level.

Generalizing this very important learning principle, all the incidents that occurred in the cluster of participating plants in the five years previous to drafting the M-PSMS are reported to the MPC. Moreover, the possible (*theoretical*) cross-company consequences of these accidents are reported. In this way, plant management is aware of the possible disastrous consequences of "simple" accidents, inducing a multi-plant-safety culture.

4.5 Practical Recommendations for Achieving Plant or Multi-Plant-*Security* Loop of Continuous Improvement

4.5.1 Introduction

To set up an effective security-management program in a chemical plant or multi-plant area, analogous to safety actions, security actions can be categorized into three main headings: *prevention of (intentional) chemical accidents, mitigation of chemical accidents,* and *follow up and corrective actions.* The security issues required for the mitigation of chemical accidents are identical to the safety issues and can be found in Section 4.4.3. In the following sections the two remaining topics to be included in the M-PSMP are discussed.

4.5.2 Prevention of (Intentional) Chemical Accidents

Threats may come in different forms and from different sources. Threats from outside the company can affect personnel and the facility itself, and may involve trespassing, unauthorized entry, theft, burglary, vandalism, bomb threats, or terrorism. Internal threats might include theft, substance abuse, sabotage, disgruntled employee or contractor actions, or workplace violence. Protective security issues can be summarized into 10 subjects. Thorough reference guidelines to these subjects are set out in the following documentation–that is Site Security Guidelines for the U.S. Chemical Industry (American Chemistry Council, 2001), Guidelines for Analyzing and Managing the Security Vulnerabilities of Fixed Chemical Sites (CCPS, 2003), General Security Risk Assessment Guidelines (ASIS, 2003), and Protection against Terrorism (MI5, 2005) – and offer recommendations for drafting a plant security-management program. To elaborate a multi-plant security-management program, some additional features are needed for each of the ten subjects. These extra multi-plant requirements are discussed in the remainder of this section.

4.5.2.1 Execution of Security-Risk Assessments (Security-Vulnerability Analyses)

The assets that need to be protected, the threats that may be posed against those assets, and the likelihood and consequences of attacks against those assets, need to be inventoried. Since the security risks and needs of individual chemical plants can vary substantially from one to the next, a chemical industrial area is unique in its approach to security-risk assessment. Obtaining a security vulnerability picture of the entire cluster of participating plants can therefore offer valuable information. The MPC uses all relevant security information from the participating plants to carry out such a security-risk assessment. Public authorities responsible for local security are informed about the potential consequences for the community of potential multi-plant-aimed attacks. To optimize integrated site security, security analyses results are discussed in group meetings with security managers from the different companies involved. Chapter 5 offers concrete suggestions on how to organize security collaboration.

4.5.2.2 Focus on Security

Security is considered at the planning stage, for example, if one of the participating plants of the initiative acquires or extends premises. Personnel and contractors working in and for the multi-plant council are thoroughly screened before employment. Relationships with law-enforcement agencies and surrounding communities are established by the MPC in order to address security concerns and to be able to intervene jointly as quickly as possible.

4.5.2.3 Security Promotion

The multi-plant council is responsible for ensuring that security is represented at a senior level in the participating organizations. The MPC also renders multi-plant security awareness part of the individual plants' safety and security cultures. To this end, the MPC periodically evaluates the effectiveness of individual companies' security-management programs. To further increase the importance of multi-plant security, posters and pictograms in the initiative's plants will indicate security as a shared responsibility of all multi-plant-related workers.

4.5.2.4 Good Basic Housekeeping

Throughout the multi-plant area's premises, public areas (whether within or outside of individual plants' fences) are kept tidy and well lit, unnecessary furniture is removed, garden areas are kept clear, and transport areas (road, railway and ship) are well maintained. The multi-plant council drafts a multi-plant housekeeping program to ensure that these basic requirements are fulfilled for multi-plant areas and to inspect the conditions in individual plants. Public authorities are responsible for non-multi-plant areas.

4.5.2.5 Reduction of Access Points to a Minimum

The multi-plant council limits the number of access points to the multi-plant area. Analogous to plant personnel and visitors, MPC staff and MPC visitors are issued with passes. Unauthorized vehicles are not allowed inside the initiative's participating plants.

4.5.2.6 Installation of Appropriate Physical Measures

Locks, alarms, control systems, and surveillance equipment are installed by the multi-plant council in agreement with individual companies at multi-plant points identified as critically vulnerable. Lighting is guaranteed throughout the multi-plant premises. A system of employee and contractor ID badges is instituted inside the entire multi-plant area.

4.5.2.7 Personnel Security

In the pre-employment phase, identities are checked and references are followed up by both the individual plant and the MPC security data administration. It is axiomatic in security that employees and contractors serve as the eyes and the ears of a multi-plant-wide security effort. The MPC security data administration therefore establishes an anonymous employee security hot line.

4.5.2.8 Enhanced IT Security Precautions

In a chemical cluster, protecting information and computer networks means more than safeguarding a single company's proprietary information (from other concurrent companies or from outsiders) and keeping the business running, important as those goals are. It also means protecting chemical processes from hazardous disruptions and preventing unwanted chemical releases with potential cross-company consequences. To an adversary, information and network access can equal the power to harm the multi-plant area, a plant belonging to the initiative, employees, and even to other plants and the community at large. Therefore, all MPC computers and the MPC electronic network are not only physically protected, but in addition software is also employed such as firewalls, virus protection, encryption, user identification, message and user authentication. A multi-plant council data administration mailroom is established away from the main multi-plant premises.

4.5.2.9 Planning and Testing of Business Continuity Plans

Business-continuity planning is obviously not just driven by security-related subjects. Nonetheless, it would be critical to the multi-plant area's survival if it were affected by a large-scale security-related incident. The MPC is responsible for drafting cross-company business-continuity plans to ensure the continuation of crucial operations should differing scenarios of knock-on accident(s) occur. These plans set out the roles and responsibilities of the different companies involved in a major (intentional) accident and of other plants not involved in the multi-plant initiative (potentially required to supply critical raw materials to companies dependent on them, *etc.*). The plans also discuss, for example, loss of staff through death or injury, a pandemic, disruption to transport, disruption to organizations on which the multi-plant area may depend, damage to multi-plant reputation, loss of IT systems, *etc.* The multi-plant council is responsible for organizing practice drills for the plans.

4.5.2.10 Documentation Control and Records

By keeping detailed records of security incidents that take place throughout the multi-plant area, the MPC monitors trends and investigates facts. To this end, a central security database to be operated by the MPC is installed. All security incident information reported to plant-security management is presented in detail to the MPC data administration security database helping to bring an offense or loss pattern to light and identify underlying issues of security concern.

Besides safety-related information, the emergency-responder tool (see Section 4.2.4.3) may also provide full integration of the geospatial and geographic information system software, rendering a predictive analytical component that creates increased capability for multi-plant intelligence and law enforcement in combating potential terrorism and conventional criminal activities in a preventive and pro-active sense.

4.5.3 Follow-Up and Corrective Actions

The follow-up of multi-plant council security recommendations and the periodic confirmation that MPC advice is up-to-date (representing current security policies, practices, procedures, and countermeasures) and is implemented at the single-plant level is an important aspect of improving overall multi-plant security.

Threats, vulnerabilities, or scenarios of particularly high risk identified by the MPC are communicated to concerned plant-security management as early as possible, preferably accompanied by countermeasure suggestions. The multi-plant council has a system for tracking the resolution and completion of the security recommendations it has offered to its participating companies. This will ensure that these recommendations are not lost or that the review and resolution of them is not delayed for unreasonable time periods.

The technical basis for the resolution of accepted and rejected MPC recommendations is documented to the MPC with a view to offering a better understanding of the reasons behind the final dispositions. This documentation may later be used by the MPC as and if required.

Since site assets and potential threats and vulnerabilities change over time as a result of changes in product line(s), customers, systems, processes, hazardous materials, *etc.*, the MPC inspects the periodical re-evaluations of plant-security management processes. Selected countermeasures are verified as being complete and as covering the current situation with respect to security in the multi-plant area.

4.6 System Implementation

After developing a plant/joint SMS or SMP, the system is implemented in the participating plant(s). Implementation procedures are drafted and responsibilities are appointed. The coordination of the development, implementation and integration of the safety-management system or the security-management program is the responsibility of the plant's department of prevention and protection at work. This department must allocate adequate and appropriate resources and personnel to each element to guarantee an effective implementation.

In the multi-plant-safety-management system and the multi-plant-security-management program, the responsibilities of the various stakeholders are documented in an implementation chapter. The chapter includes topics such as procedures for the different companies to cooperate and procedures for sharing different levels of information. This phase is carefully worked out by the MPC in cooperation with all the relevant stakeholders.

Implementation practices are checked on a regular basis and penalties may be considered appropriate if a company or a person ignores its or his task. Promotion measures to motivate companies and to stimulate implementation of the safety rules and regulations and the security recommendations and countermeasures may also be suggested.

To guarantee the success of a multi-plant-safety-management system and/or a multi-plant-security-management program, commitment of the participating companies is essential (CCPS, 1994b; CCPS, 2003). Commitment begins with the engagement of plant top-level management; the latter must be convinced of some major benefits arising from cooperation with neighboring companies. These benefits include:

- improved relationship with neighboring companies;
- an opportunity to learn from the experiences of other companies to improve safety and security know-how;
- improved image within society and with one's own employees;
- improved position for negotiations with the insurance industry;
- decreased probability of escalation events;
- better coordination in the event of major cross-company (accidental or intentional) accidents;
- more justified legal conditions in the event of a major domino event occurring;
- increased efficiency of allocating security measures;
- huge hypothetical financial benefits (see section 2.7);
- possible immediate (concrete, non-hypothetical) financial benefits (see Chapter 9).

Support of lower management and employees is also needed. However, if top management is prepared to cooperate and to take the necessary actions, it is most likely that line management will follow suit.

Commitment should be voluntary. Company top management should have the "foresight" to recognize the importance of multi-plant safety and security. For organizations to be successful, dealing with the future should be a background organizational skill. In executing its primary task, processing, storing or transporting hazardous substances, a chemical company acts necessarily in the present. To be able, however, to continue processing hazardous substances, it needs to be concerned not just with the present but with the future as well. In dangerous high-tech chemical surroundings, safety and security management and joint safety and security management become key plant concepts for global strategic sustainability. However, joint safety and security management in particular has not yet received the attention nor the level of importance it deserves.

When expanding a plant-safety-management system to a multi-plant-safety-management system or indeed a plant-security-management program to a multi-plant-security-management program, a second emphasis lies on the communication between the companies involved. The participating companies share certain information to a certain degree, such as the kind of hazardous chemicals used in their various production processes, the location of these chemicals, changes made to processes in particular those having consequences for the other participating plants, regular reviews of the effectiveness of the system via communication between plant-safety managers, *etc.* Moreover, communication between the employees of the different participating companies is encouraged with a view to sharing experiences and learning from other plants.

Multi-plant safety and security communication can be divided into two levels, that is the department level and the (aggregate) multi-plant level. The highest level of S&S communication is needed by the department level, as cross-department meetings lead to a reciprocal type of communication. The structure of the departments allow for frequent horizontal communication, perhaps through the use of permanent teams. Weekly interaction between departments is required and department managers are jointly involved in face-to-face communication, information exchange and decision making.

On a multi-plant level, communication is sequential by nature. Once the output of the departments has been given to the respective MPC department think tanks, these think tanks have to formulate recommendations and guidelines that can serve as input for the think tank of the safety management department and the security management department. The latter departments transform the various inputs into procedures and rules and standardize the multi-plant-safety-management system and multi-plant-security-management program. In addition, typical multi-plant safety and security issues are also added by safety- and security-management departments and independent experts. The adjusted M-PSMS and M-PSMP documents as a whole are then used as a feedback input to the different departments.

4.7 Summary and Conclusions

This chapter addresses the design, development, and organization of multi-plant-safety-management systems and of multi-plant-security-management programs. Its objective is to offer guidance for installing a workable, effective multi-plant management program, fully integrated into the individual company's safety and security-management program(s), and respecting its business priorities, culture, and organization. For this reason, the chapter does not present formulae nor does it dictate instructions for implementation. Instead, it aims at providing information and suggestions to help chemical plants participating in a multi-plant initiative take up their responsibilities and jointly develop a multi-plant-safety-management system as well as a multi-plant-security-management program, and thereby achieving higher safety and security levels within the multi-plant area at acceptable costs.

When drafting multi-plant safety and security procedures or when planning risk analyses to avoid external domino effects, company confidential information has to be used. To this end, chemical plants should jointly establish a body at site-integrated (multi-plant) level, the multi-plant council, responsible for multi-plant safety and security topics and (to be workable) divided into two parts: a part comprising multiple company staff and a part comprising independent personnel handling confidential data. By implementing the suggested approach in this chapter and establishing such a site-integrated body composed of a "confidential part" and a "non-confidential part", the "lack of openness" – problem (note that openness is required for successful collaboration) is taken into consideration and solved. Using the proposed multi-plant approach also leads to safety and security improvements of companies situated within the same (large) industrial area (e.g., the Houston chemical cluster), but, for example, not situated close to each other. It also enhances – to a lesser extent and in an indirect way – safety and security levels in plants not participating to the multi-plant initiative.

5
A Multi-Plant Safety and Security Culture – The People: Facilitating Multi-Plant Safety and Security Collaboration

5.1
Introduction

As mentioned in Chapter 2, from the point of view of a company, domino-effect risks can be internal or external in nature. Internal domino effects completely unfold on the enterprise premises, while external domino effects are characterized by multi-plant involvement. In this book we focus on improving the prevention of the latter type of domino effects. It is crucial to know and to minimize the pathways that lead to such multi-plant knock-on effects. However, despite the fact that external domino effects are important hazards in the chemical industry, possibly leading to disaster, risk prevention on these effects has not been a priority of multi-company policy. Various reasons such as multi-organizational learning barriers and the need for intensive cooperation between two or more often competing companies can explain this behavior. Thus, despite the fact that external domino-prevention management within a chemical cluster should be the shared responsibility of all chemical plants concerned, external domino-safety risks are hardly or not explored at company level or at multi-company level. In fact, company research on external domino effects is limited to the compulsory exchange (if applicable) of information on hazard- and system characteristics with adjacent companies. Nonetheless, chemical plants should join forces to elaborate a risk-analysis-management technique or framework specifically aimed at dealing with external domino risks from a safety as well as a security point of view.

In general, successful safety-risk management is based on a comprehensive and detailed hazard mapping and a full understanding of possible accident consequences (Wells, 1997). Risk-analysis methods should be chosen carefully based on strengths of each technique, as each may provide different outputs leading to different prevention measures. Hence, there is a need for the development of an external domino-accident-prevention framework integrating well-known risk-analysis techniques with cross-company cooperation recommendations. Also, specific techniques, developed by experts of the company itself and with a lot of company know-how, remain indispensable for a firm safety management. Companies are recommended to use common safety-analysis tools for intra-company use to enable future inter-company cooperation. Widely known and accepted techniques

Multi-Plant Safety and Security Management in the Chemical and Process Industries. G.L.L. Reniers

ISBN: 978-3-527-32551-1

can be standardized and optimized and their implementation can be verified. Through the exchange of expertise, cooperation will lead to a safer working environment.

Furthermore, a framework and a network for enhancing cross-company security-management collaboration should be elaborated as well. Security management from different plants should be able to cooperate at two levels: "indoor" and "outdoor".

Carrying out security-vulnerability assessments (SVAs) has only gained major significance in chemical firms relatively recently, and as a consequence the SVA procedure has a standardized character. A lack of variety in security-vulnerability-assessment approaches renders collaboration between plant indoor security management and indoor security managers belonging to other plants significantly easier. Nonetheless, a framework assisting collaboration between indoor security managers belonging to a group of companies as they carry out multi-plant security-vulnerability assessments (SVAs) needs to be elaborated.

Furthermore, elaboration of a security network is required as a way of rendering collaboration between outdoor security guard services of neighboring plants more attractive.

A security manning level evaluation instrument should be envisioned as well. Appraising the quality and quantity of security staffing levels in industrial areas can prove to be decisive in preventing, for example, terrorist attacks. At all times, under any circumstances, and at locations throughout the industrial area, critical infrastructure needs to be protected. A lack of surveillance or of qualified security personnel can lead to a substantial increase in asset vulnerability. Therefore, a method to help evaluate whether the industrial area has sufficient security personnel is demanded. The method should be usable by individual organizations as well as by groups of organizations.

5.2 A Multi-Plant-*Safety*-Management Framework

5.2.1 Inherent Safety

The best way to achieve safe plant operation is to have inherently safe processes, and to operate in correctly designed, controlled and maintained installations equipment. However, it seems inevitable that once process and plant designs have been engineered to optimize processing safety, a spectrum of operating safety risks remain in many chemical operations. To deal with these safety risks, a comprehensive plant safety-management system is developed (see Section 4.2.2). The PSMS addresses process-hazard assessment, specification of safety risk-control measures, evaluation of the consequence of failures of these controls, documentation of engineering controls, and scheduled maintenance to assure the on-going integrity of the protective equipment. Basic process control and safety instrumented functions (SIF[1]) must be considered in each of these segments. A SIF is a combination of sensors, logic solvers, and final elements with a specified safety integrity level (SIL) that detects an out-of-limit condition and brings the process to a functionally safe state. However, these methodologies do not investigate the underlying physical and chemical hazards that must be contained and controlled for the process to operate safely, and thus they do not integrate inherent safety into the process of achieving safe plant operation. Most action items refer to existing safety procedures or refer to technical safeguards or require the addition of new levels of protection around the same underlying hazards. However, taking preventive measures in the conceptual design phase is extremely important to help achieving plant safety and multi-plant safety.

The most desirable requirement of chemical equipment is that it is *inherently safe*. Achieving such inherent safety starts in the process-design phase of the equipment. An inherently safe process-design approach includes the selection of the process itself, site selection and decisions on dangerous substances inventories and plant layout. Complete inherent safety is rarely achievable within economic constraints. Therefore potential hazards remaining after applying such an approach should be addressed by further specifying independent protection layers to reduce the operating risks to an acceptable level.

In current industry practice, chemical facilities processing dangerous substances are designed with multiple layers of protection, each designed to prevent or mitigate an undesirable event (Thurston, 1994). Multiple, independent protection layers (IPL) addressing the same event are often necessary to achieve sufficiently high levels of certainty that protection will be available when needed. Powell (1996) defines an IPL as having the following characteristics: (i) *Specific*–designed to prevent or to mitigate specific, potentially hazardous events, such as a runaway reaction, release of toxic materials, loss of containment or fire; (ii) *Independent*–independent of the other protective layers associated with the identified hazard; (iii) *Dependable*–can be counted on to operate in a prescribed manner with an acceptable reliability. Both random and systematic failure modes are addressed in the assessment of dependability; (iv) *Auditable*–designed to facilitate regular validation (including testing) and maintenance of the protective functions; (v) *Reducing*–the likelihood of the identified hazardous event must be reduced by a factor of at least 100.

An IPL can thus be defined as a device, system or action that is capable of preventing a scenario from proceeding to its undesired consequence independent of the initiating event or the action of any other layer of protection associated with the scenario.

Figure 5.1 illustrates safety layers of protection used in the chemical industry. Detailed process design provides the first layer of protection. Next come the automatic regulation of the process heat and material flows and the provision of sufficient data for operator supervision, together called the "basic process control systems" (BPCS). A further layer of protection is provided by a high-priority alarm

1) This is a newer, more precise term for a safety interlock system (SIS).

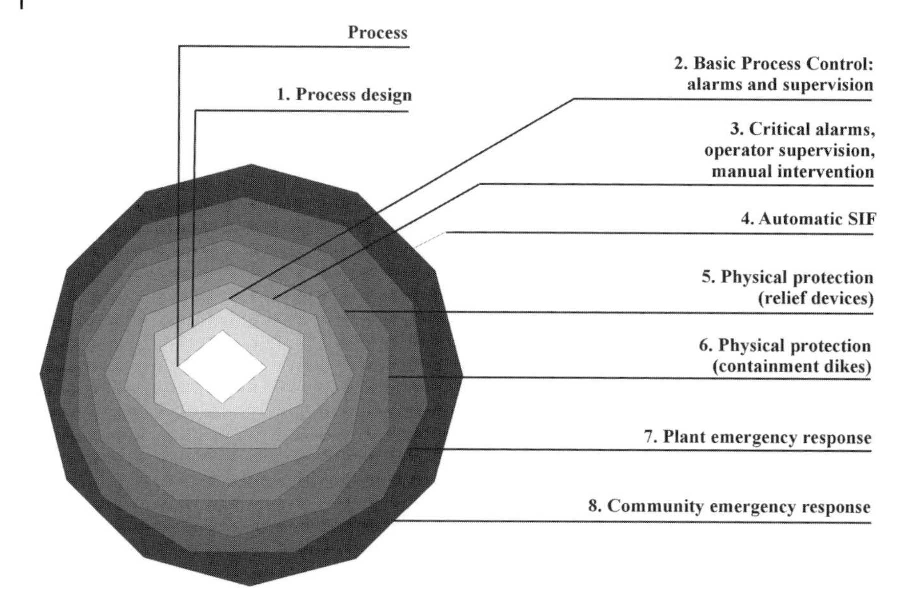

Figure 5.1 Typical layers of protection found in modern chemical plants (Source: based on CCPS, 1993).

system and instrumentation that facilitates operator-initiated corrective actions. A safety instrumented function, sometimes also called the emergency shutdown system, may be provided as the fourth protective layer. The SIFs are protective systems that are only needed on those rare occasions when normal process controls are inadequate to keep the process within acceptable bounds. Any SIF will qualify as one IPL. Physical protection may be incorporated as the next layer of protection by using venting devices to prevent equipment failure from overpressure. Should these IPL fail to function, walls or dikes may be present to contain liquid spills. Plant and community emergency-response plans further address the hazardous event.

By considering the sequence of events that might lead to a potential incident, another representation can be developed highlighting the efficiency of the protection layers, as shown in Figure 5.2.

Figure 5.2 illustrates the benefits of a safety risk-identification approach to examine inherent safety by considering the underlying hazards of a hazardous substance, a process or an operation. Inherently safer features in a process design can reduce the required SIL (see Table 5.1) of the SIF, or can even eliminate the need for a SIF, thus reducing cost of installation and maintenance. Indeed, the Center for Chemical Process Safety (CCPS, 1996) suggests that added-on barriers applied in non-inherently safer processing conditions have some major disadvantages, such as the barriers are expensive to design, build and maintain, the hazard is still present in the process, and the accumulated failures of IPLs still hold the potential to result in an incident.

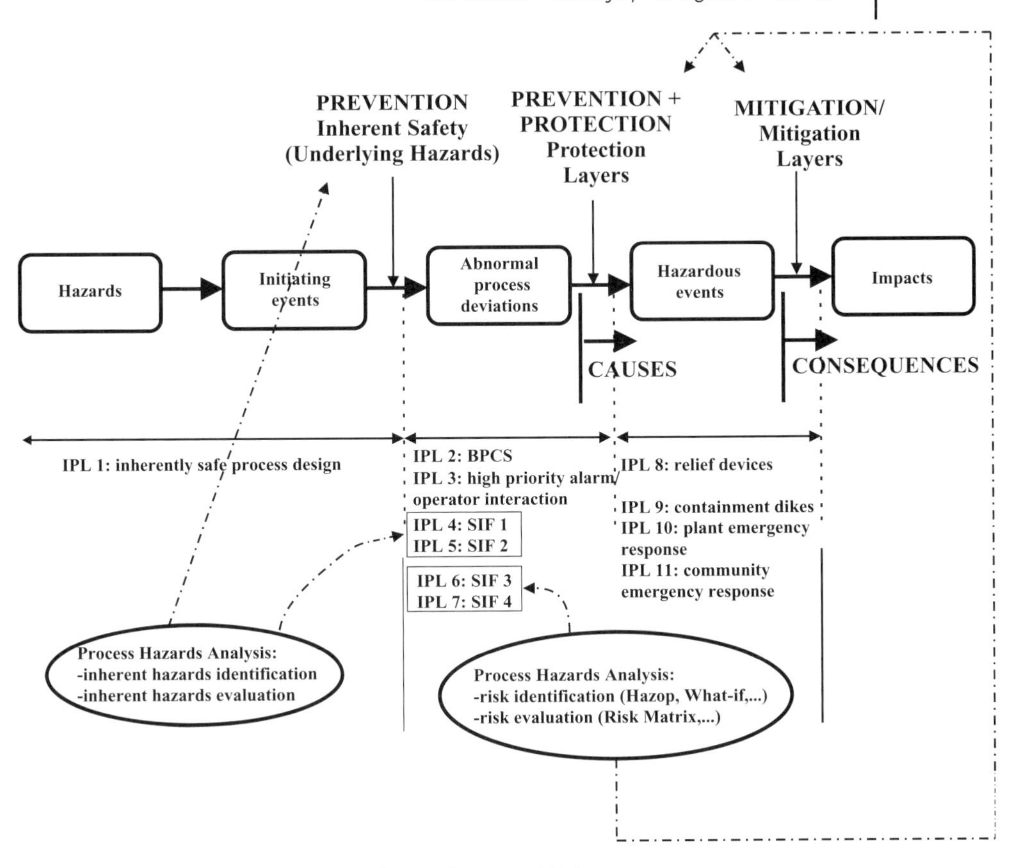

Figure 5.2 Anatomy of a chemical incident (with, e.g., 4 SIFs).

Table 5.1 Safety integrity levels.

Safety integrity level (SIL)	Safety availability	Probability to fail on demand (PFD)	Equivalent risk reduction factor (1/PFD)
SIL 4	>99.99%	≥10–5 to <10–4	10 000–100 000
SIL 3	99.9–99.99%	≥10–4 to <10–3	1 000–10 000
SIL 2	99–99.9%	≥10–3 to <10–2	100–1 000
SIL 1	90–99%	≥10–2 to <10–1	10–100

Source: based on Gruhn, 1999 and International Standard, 2003.

The primary purpose of evaluation methods currently employed throughout the chemical industry, such as the risk matrix or layer of protection analysis (LOPA), is to determine if there are sufficient layers of protection against an accident scenario. A scenario may require one or more protection layers depending on the process complexity and the potential severity of a consequence. Note that for a given scenario only one layer must work successfully for the consequence to be prevented. However, since no layer is perfectly effective, sufficient protection layers must be provided to render the risk of the accident tolerable. Therefore, it is very important that a consistent basis is provided for judging whether there are sufficient IPLs to control the safety risk of an accident for a given scenario. Especially in the process-design phase an approach for drafting inherent safety into the process by implementing satisfactory IPLs is needed for effective process safety. In many cases the safety instrumented functions are the final independent layers of protection for preventing hazardous events. Moreover, all SIFs are required to be designed such that they achieve a specified safety integrity level (SIL).

The SIL is the quantification of the probability of failure on demand of a SIF into four discrete categories. Table 5.1 gives an overview of such levels (as already mentioned).

Thus, four corresponding degrees of reduction in hazardous event likelihood are produced by the SILs. SIL 1 provides about two orders of magnitude of event likelihood reduction; SIL 2 about three orders of magnitude, SIL 3 about four orders of magnitude and SIL 4 more than four orders of magnitude. Obviously, the availability targets for SIL 3 and SIL 4 are extremely stringent and the design practices to achieve and maintain these high levels are extensive and costly.

Gardner and Reyne (1994) point out that methods used to select safety integrity levels are based on an evaluation of three characteristics of the process and the hazardous event associated with the SIF: the severity of the hazardous event consequences (minor, serious, extensive), the likelihood that an upset situation will occur that could lead to these consequences (low, moderate, high) and the number of IPLs. Before a SIL can be selected, the inherent risk of the process must be evaluated. Next, credit for all non-SIF mitigation measures (e.g., relief valves, dikes) must be accounted for, to determine the baseline risk of the process, which is the starting point of the SIL selection. All of the SIF design, operation, and maintenance choices have to be verified against the target SIL. The safety design engineer has to realize that further mitigation with a SIF solely reduces the likelihood of an incident. For example, if the baseline likelihood is 10^{-2} per year, a SIL 2 would reduce the likelihood up to 10^{-5} per year. The risk-reduction process is illustrated in Figure 5.3, in which risk criteria are represented in the form of FN_F limit lines.

Determining the necessity of IPLs and SIFs and the required level of their safety integrity is performed using safety risk-identification and -evaluation methods. No one approach for the selection of a SIL is appropriate in every situation.

Summarizing, the SIL must be chosen to reduce the incident frequency to a tolerable level. It is the design basis for all engineering decisions related to the safety instrumented function. When the design is complete, it must be validated against the SIL. Therefore, the safety integrity level closes the design cycle: starting with hazards identification, pursuing with requirements quantification and ending with design validation.

Achieving inherent safety is not always possible. When processing, storing or transporting hazardous materials, residual risks often remain to a greater or lesser

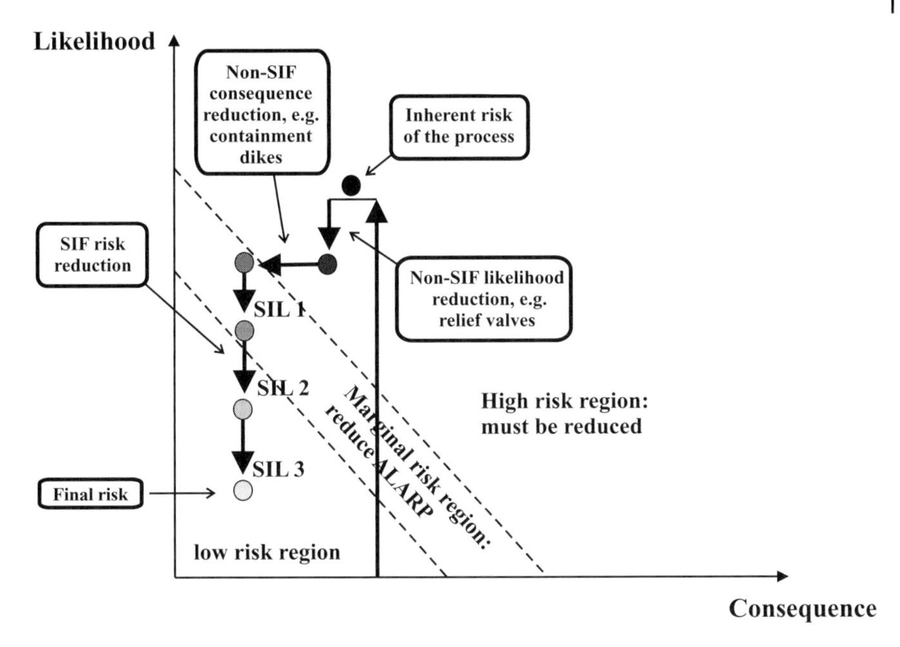

Figure 5.3 The effect of risk-reduction measures (Source: based on Marszal *et al.* 1999).

extent. To identify and to assess cross-plant residual risks in an optimal and effective way, the analyses aimed at remaining multi-plant risks should be executed by plants intensively collaborating with each other. The next sections discuss how to efficiently plan and perform process risk-analysis procedures in a multi-plant area.

5.2.2
Developing an External Domino-Accident-Prevention Framework: Hazwim

5.2.2.1 Introduction

In general, the design of precaution and protection measures at a chemical site can be considered to be a two-step risk-analysis procedure (Sinnott, 1996). The first step involves a risk analysis, the systematic examination of all possible safety risks. The second step is to evaluate these safety risks and to make a classification. Both hazard identification and safety-risk evaluation have become very important research topics. Developing insights on these subjects force companies to continuously adapt or change company safety management to improve plant safety, as indicated by Baram (1998).

Further optimization of risk-analysis methodologies beyond the current state-of-the-art becomes ever more difficult due to multi-causal dependencies and non-linearities especially in the case of major accidents. Nevertheless, because of the possibility of major accidents occurring, effective safety management is of huge importance to the chemical industry.

Empirical research expounded by Reniers *et al.* (2005a) reveals that safety managers acknowledge the importance of cross-company cooperation for reducing external domino risks. The lack of a general framework for joint external accident prevention and the fear of high joint risk-inspection and risk-analysis costs, are reported to be the main objections against inter-company safety cooperation. The complementary risk-analysis procedures of hazop (15%), the risk matrix (13%) and the what-if analysis (8%) are reported to be the most frequently used risk-analysis techniques at hazardous chemical facilities. The next section discusses these risk analyses methods more in depth. Ideally, the framework to be developed integrates these three techniques into an effective standardized risk-analysis framework for the prevention of external domino accidents in an industrial area. The combination of techniques on the one hand and qualitative and quantitative data on the other, would offer a comprehensive up-to-date list of cross-company domino hazards and recommended actions in the area under consideration. The framework offers support to prevention managers and safety policy makers concerning external domino prevention.

5.2.2.2 Hazop, What-If Analysis and the Risk Matrix

Hazop (Hazard and Operability Study) Hazop is a formalized methodology to identify and to document hazards[2] through imaginative thinking. It involves a very systematic examination of design documents that describe the installation or the facility under investigation. The study is performed by a multidisciplinary team analytically examining design intent deviations. This qualitative method gives the team members the opportunity to be creative in a systematic way (Early and Sherrod, 1991). The hazop study aims at thoroughly documenting the hazard causes[3] and hazard consequences[4] of the operation deviations and formulating recommended actions. For more interested information on carrying out a hazop, the interested reader is, for example, referred to Greenberg and Cramer (1991) or CCPS (1992).

What-If Analysis Similar to hazop, what-if analysis is a qualitative technique utilizing process-typical information for safety-risk identification. The technique is a structured brainstorming method determining what things can go wrong (using "What if" type of questions) and judging the consequences of those situations occurring. An experienced what-if review team can effectively and productively

2) A *hazard* is any unsafe condition or potential source of an undesirable event with potential for harm or damage (see Chapter 2), for example, the amount of product *A* present in a reactor exceeds design specifications).

3) A *hazard cause* is any possible cause of a hazard (e.g., the valve for product *A* fails to close).

4) A *hazard consequence* is a combination of a hazard on the one hand and a person or event on the other hand starting a chain of occurrences. The cascade begins with the presence of a hazard and ends with the consequences of an accident (e.g., some product A is unreacted and is released to the work area).

discern major issues concerning a process or system. For more interested information on carrying out a what-if analysis, the interested reader is, for example, referred to Greenberg and Cramer (1991) or CCPS (1992).

The Risk Matrix The "risk-assessment-decision matrix", often abbreviated as "the risk matrix", is a systematic approach for estimating and evaluating safety risks. This tool can be employed to measure and categorize safety risks on an informed judgment basis as to both probability and consequence and as to relative importance. An example of the risk matrix is illustrated in Table 5.2.

Once the hazards have been identified, the question of assigning severity and probability ratings must be addressed. A common, basic example of assigned ratings by a team on a generalized basis can be found in Table 5.3.

The probability level F, "impossible", makes it possible to assess residual risks for cases in which the hazard is designed out of the system.

The rankings provide a quick and simple priority sorting method. The rankings are then given definitions that include, for example, definitions and recommended actions similar to those in Table 5.4.

Such a safety-risk assessment methodology is especially implemented in the case of low-probability, high-severity hazards. Therefore, process hazards are evaluated using the risk matrix.

It is very important that frequency estimates and consequence estimates are very well considered and carried out by experienced risk managers. For more information on using a risk matrix to prioritize risks, the interested reader is, for example, referred to Greenberg and Cramer (1991) or CCPS (1992).

Table 5.2 The risk-assessment-decision matrix.

Severity of consequences	Probability of hazard					
	F	E	D	C	B	A
	Impossible	Improbable	Remote	Occasional	Probable	Frequent
I Catastrophic					1.	
II Critical				3.	2.	
III Marginal			4.			
IV Negligible						

Risk code/ actions	1. Un-acceptable	2. Un-desirable	3. Acceptable with controls	4. Acceptable

Source: based on Department of Defence, 2000.

Table 5.3 Criticality and frequency rating for the risk-assessment-decision matrix.

Severity of consequences – ratings

Category:	Descriptive Word:	Results in either:
I	Catastrophe	• An on-site or an off-site death • Damage and production loss greater than 750 000 €
II	Critical	• Multiple injuries • Damage and production loss between 75 000 € and 750 000 €
III	Marginal	• A single injury • Damage and production loss between 7 500 € and 75 000 €
IV	Negligible	• No injuries • Damage and production loss less than 7 500 €

Hazard probability – ratings

Level:	Descriptive word:	Definition:
A.	Frequent	Occurs more than once per year
B.	Probable	Occurs between 1 and 10 years
C.	Occasional	Occurs between 10 and 100 years
D.	Remote	Occurs between 100 and 10 000 years
E.	Improbable	Occurs less often than once per 10 000 years
F.	Impossible	Physically impossible to occur

Source: based on Department of Defence, 2000.

Table 5.4 Definitions and recommended actions for rankings.

Ranking	Description	Required action
1.	Unacceptable	Should be mitigated with technical measures or management procedures to a risk ranking of three or less within a specified time period such as, for example, six months.
2.	Undesirable	Should be mitigated with technical measures or management procedures to a risk ranking of three or less within a specified time period such as, for example, twelve months.
3.	Acceptable with controls	Should be verified that procedures or measures are in place.
4.	Acceptable	No mitigation action required.

Risk-Analysis Techniques: Summary Developing an integrated risk-analysis scheme that combines a well-considered selection of widespread techniques would enable hazardous chemical companies to achieve more efficient joint safety-risk identification and could lead to joint domino safety plans in the long run. Hazop is an expensive, complex technique demanding a lot of expertise and implementation time offering very detailed safety-related documentation. A high-frequency application of this method is obviously not optimal. The what-if methodology on the other hand is one of the simplest forms of conducting a hazard analysis. Moreover, this low-cost technique does not require special quantitative methods or extensive preplanning. A combination of both these methods seems to be an economically optimizing alternative. In practice, the hazop and what-if safety-risk-identification techniques have much in common, and a meta-technical complementary combination for external domino-hazards assessments offers cooperation opportunities for all the participating companies. The well-known and user-friendly safety-risk evaluation technique of the risk matrix is the most applied safety-risk-ranking method in the chemical industry and can thus be used to evaluate hazop results as well as what-if results.

5.2.2.3 An External Domino-Accident-Prevention Framework: Hazwim

Cross-Company Management Cross-company management essentially differs from single-company management by the amount of information managers have at their disposal. Another problem is that different organizational perspectives have to be combined. As Hovden (1998) explains, each frame or perspective provides a different way of interpreting events and actions, and each implies a different focus with consequences for choice of strategies and approach to effective management.

Hence, bringing these viewpoints together is not an easy assignment. There should be no communication problems or misunderstandings between the parties involved. Therefore, a safety-risk-management strategy has to be worked out to be sure that responsible personnel from different plants communicate with the same know-how, on the same level, and about the same safety issues. For this purpose, a cross-company safety-risk-management overview based on a single-company model from the Australian/New Zealand Standard (Standards Australia/Standards New Zealand, 1999) can be developed. The safety-risk-management iterative process is illustrated in Figure 5.4.

This overview can be used for developing a framework that supports inter-company information exchange on external domino effects. To provide a situation where communication conflicts are minimized, the risk-analysis procedures and their results need to be understood by the different experts concerned with the study. In the next section, a framework is elaborated based on the cross-company risk-management overview of Figure 5.4.

The Hazwim Framework The aim of an external domino-accident-prevention framework is to facilitate the structuring of off-site domino-prevention cooperation

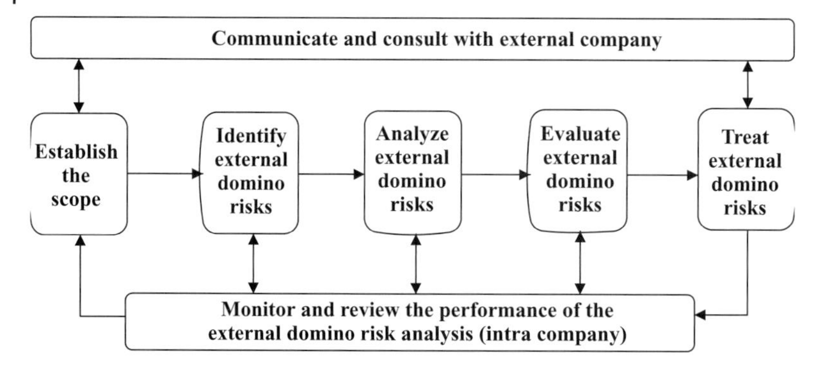

Figure 5.4 Cross-company safety-risk-management overview.

between plants handling or storing hazardous substances. The framework should include a solid external domino-risk analysis consisting of two stages providing key information for external domino-risk management: external domino-hazard identification and external domino-risk evaluation.

Combining hazop analysis, what-if analysis and the risk matrix into one hazwim framework allows for constituting a meta-technical tool for optimizing the organization of discussing process-hazard analysis performances by employees of neighboring companies in an industrial area.

Prior to the first stage of the hazwim framework, hazwim facilitators and experienced project managers have to be appointed in every company. In the first stage of the framework (see Figure 5.5), the scope of the external domino effects study is defined. The scope area, ideally being a sub-cluster that forms an island where no domino effects can enter or leave the area, results from a technical security risk related domino effects investigation (see Chapter 6) carried out by the MPC data administration (see Chapter 4). Afterwards, the facilitators and the project managers in every company further refine the scope of the study and the scope area. In this way, every domino island can be sub-divided into several scope areas. These scope areas are ranked and hazwim proceeds with the first priority scope area. In the second step, a technical study is performed by experts from the MPC data administration to identify the potential per installation (being part of the scope area) for continuing domino effects and their consequences. Hence, a prioritization is made and the installations are ranked in order of decreasing danger (see Chapter 6 for information on the technology and tools required to carry out this exercise). The study results are communicated to each company being part of the scope area. In the next step, the company project engineer sets up the organization schedule for the scope area of the own company using the classification data and, if it is not the first time the installation is under consideration, the previously used hazwim scheme for the installation (to decide whether a hazop or a what-if analysis is needed). Further, domino-hazop and domino-what-if teams are selected in each company (see Section 6.2 at the end, for more information on executing a domino-hazop). Once the teams are selected, a guided tour of

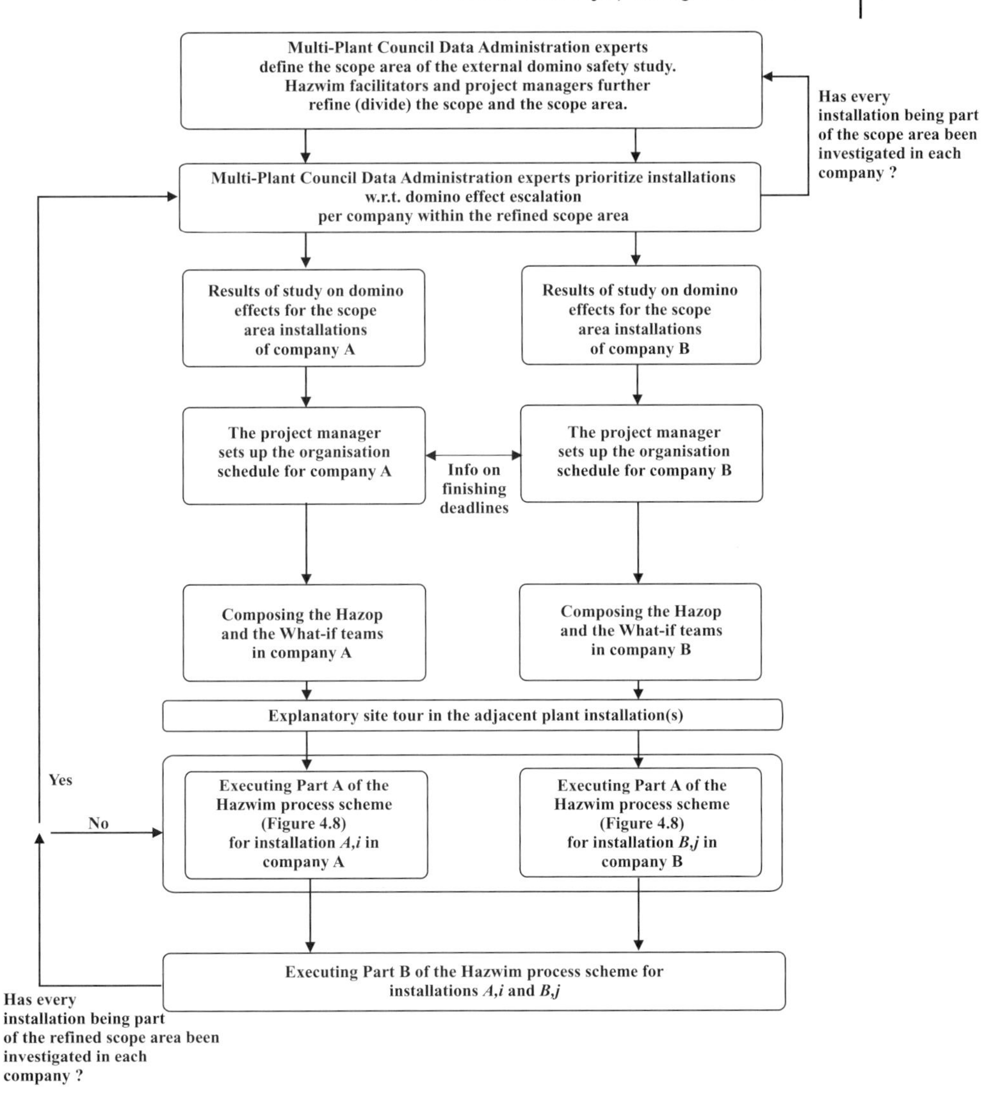

Figure 5.5 The hazwim framework.

the scope area of the adjacent companies is organized. The hazwim project participants then execute stage seven of the hazwim framework. This final step of the framework cycle is completed when every installation that has been retained to investigate for domino-effects prevention in the scope area under consideration, has been checked by both parts of the hazwim process scheme (see Figure 5.8). Afterwards, together with the project managers the hazwim facilitators define a new scope for a new area. A thorough elaboration of the different framework stages is given in the following sections: 5.2.2 Step 1 to 5.2.2 Step 7.

Step 1: Defining the Scope and the Scope Area of an Industrial Area External Domino Safety Study Successfully implementing the domino-risk-analysis framework depends on effectiveness, efficiency, and "managing the manageable". The physical boundaries of the hazwim study are made explicit by MPC data administration specialists using a computer-automated tool (for development recommendations, see Chapter 6). Furthermore, the level of detail of the hazop and the what-if analyses to be carried out should remain within reasonable boundaries defined by the level of formal control that safety management needs to exert. The scope of the study should therefore describe the types of hazards and hazard controls in the different installations to be covered by the hazop or the what-if team. To this end, hazwim facilitators and project managers collaborate to make alterations to the scope area according to their joint insights. At the end of this stage, concretized scope area identification can be expected, as depicted in Figure 5.6.

Step 2: Prioritizing Installations Each installation in an industrial area is a potential threat to its environment. The whole installation area should therefore be considered in the analysis. The implementation of each hazwim scheme is influ-

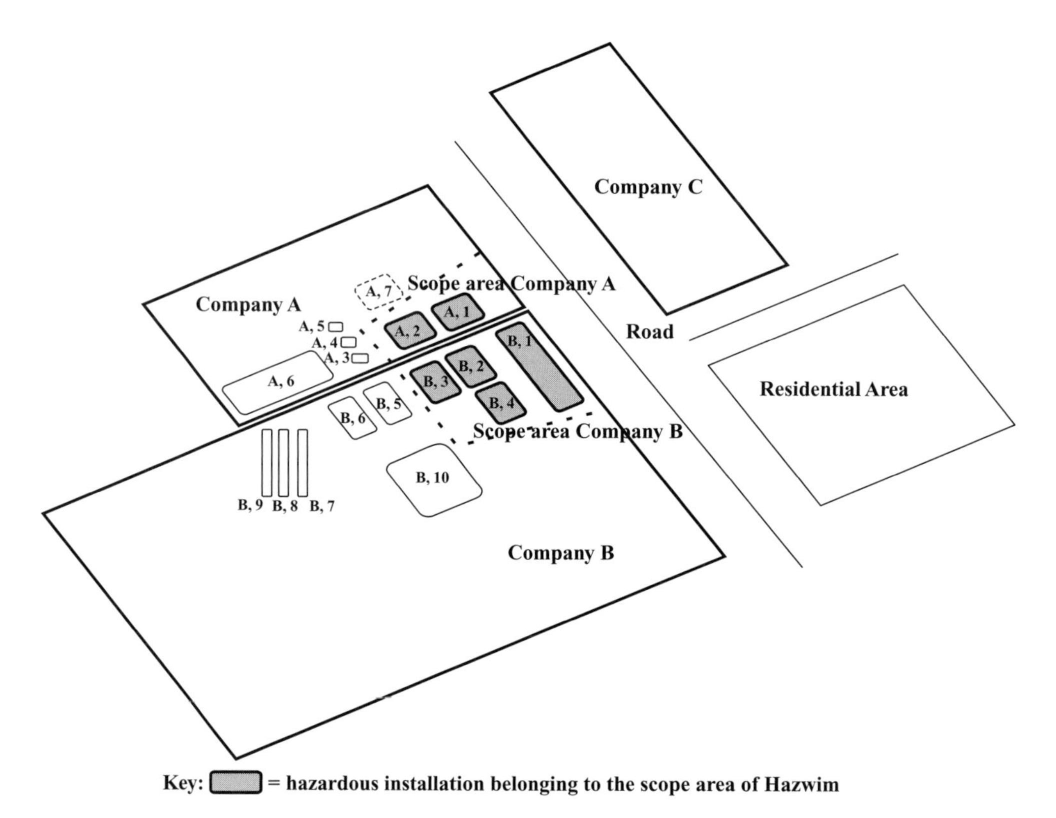

Figure 5.6 Identifying the scope area in the hazwim framework for two fictive neighboring companies *A* and *B*.

enced by danger correlations between the installations' process hazards on the one hand and another installation, a process unit or a collection of process units on the other. Potential interactions between parts of units or facilities, which may not be hazardous in themselves but that may cause some other parts of the plant to perform abnormally and result in an accident sequence, may also exist.

In chemical enterprises where there are numerous hazards with severe potential consequences, computer programs are currently used to calculate the safety risk levels on a topological grid, and then to plot contours of safety risk on the grid. These contours are used to display the frequency of exceeding excessive levels of hazardous exposure. However, available software packages fail to offer a straightforward prioritization of installations within an extensive industrial area with respect to continuing domino effects. For this purpose, Chapter 6 suggests how to design and elaborate (external) domino software and elaborates its concepts and implementation strengths and weaknesses in detail. In this stage, the refined scope area (i.e. a domino island or a part of it) is investigated by MPC data administration specialists tabulating a ranking of installations contributing to domino-effect escalation.

Step 3: Sharing Study Results with Companies The results of the study performed by the MPC data administration are communicated to the companies concerned. If required, an explanation of the results is offered to the companies' safety teams and project managers.

Step 4: The Organization Schedule The need for flexibility in applying hazwim schemes on various installations and the involvement of different plants lead to a complex situation in which timing requirements, responsibilities and competences must be well defined.

Each plant's hazwim project manager chronologically plans the various hazwim process scheme steps for the plant's installations situated in the area under consideration.

Installations in the area to be checked are listed and for every installation an overview table is made. This overview table contains at least the relative danger level of each installation and the time needed to perform a hazop or a what-if procedure on the installation. The relative danger level (obtained via the MPC study) is complemented with safety-risk contours information. Safety-risk contours are calculated by plant safety engineers using specific risk-analysis software tools.

Based on these installation data, an optimizing time schedule to investigate the external domino danger in the scope area can be drawn. An example of such a time schedule is given in Figure 5.7.

In the example, a hazop is first carried out on installation *A, 2* in company *A*. Sometime later, a what-if analysis is executed on installation *A, 1*. In company *B*, a what-if analysis on *B, 1* is followed by a hazop on *B, 3* and on *B, 2* and finally a what-if performance on *B, 4*. The schedule is executed according to the hazwim process scheme from Section 5.2.2 Step 7.

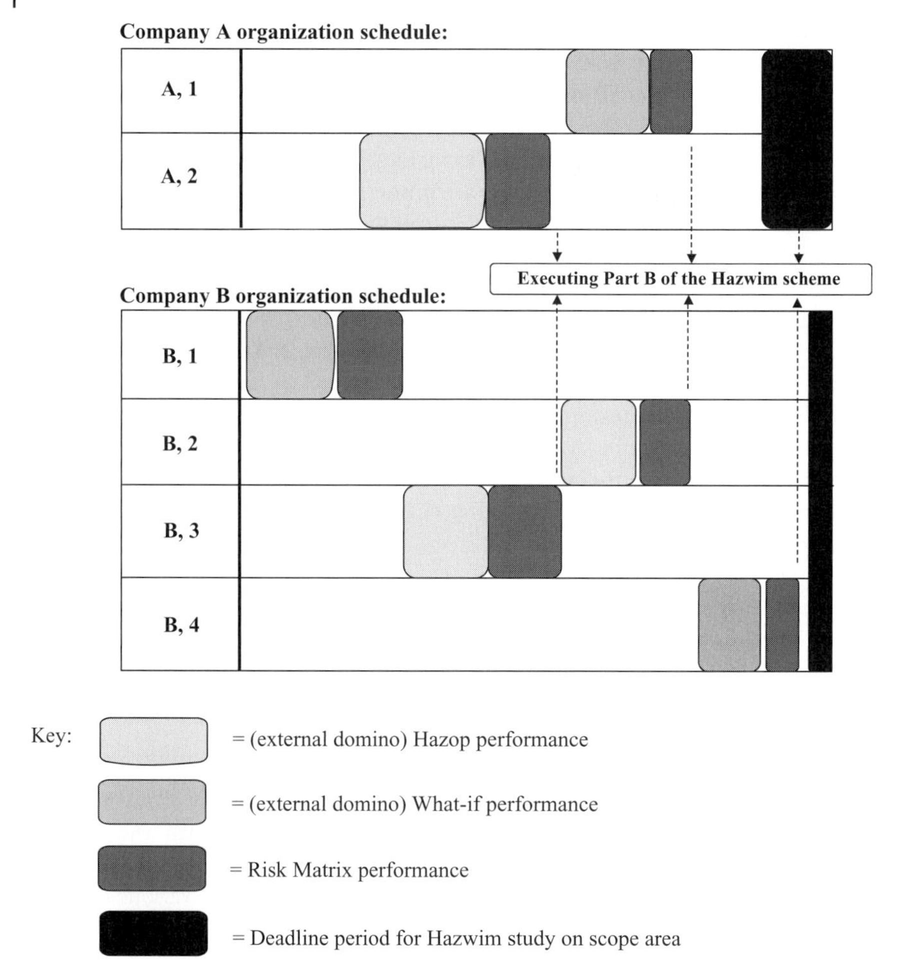

Figure 5.7 Hazwim organization schedule for two fictive neighboring companies A and B.

Information on finishing deadlines for all hazwim performances in both companies is exchanged.

Step 5: Composing the Hazop and What-If Teams in Each Company The key to success (and efficiency) in executing a solid hazwim framework is the presence of appropriate skills of team members. Hazwim facilitators should be familiar with hazop and what-if analyses as well as with the risk matrix evaluation method. They should also have process and maintenance knowledge of the installations situated on the premises of their companies. Part-time members, such as process engineers and maintenance engineers from the installations to be reviewed should remain on call for information assistance.

Different situations have to be considered when composing a suitable hazop/what-if team. Per company two safety-risk-identification procedures have to be carried out, so theoretically two teams are needed. Because the procedures are performed at different moments in time and the two teams also possess approximately the same characteristics and functions, the same team can act as a hazop team and, reduced, as a what-if team. In this way, the two types of studies are conducted mainly by the same experienced personnel. Only then will the imaginative character of the what-if study be at full strength and the least number of hazards will be overlooked. The hazwim facilitator from a certain company is also the hazop/what-if team leader from that company. The teams can be extended by specialists depending on the objectives and boundaries of the study and on the installations' characteristics. The size of the hazop/what-if team is recommended to be about five to nine persons depending on the study objectives and boundaries.

Step 6: Explanatory Site Tour In hazwim an elementary review of the technical characteristics of neighboring installations is necessary as the focus will be on hazards that can lead to off-site domino accidents. The review will allow the team members to imagine possible deviations, hypotheses and questions on the operation of the examined facility and adjacent facilities. Therefore, an explanatory site tour in the adjacent plant installations is indispensable. Once all team members are chosen, such a guided tour can be organized. It will support the collection of installation-specific information on risk contours and dangerous substances.

Step 7: The Hazwim Process Scheme The hazwim process scheme combines the safety-risk-identification techniques of hazop and what-if analysis with the risk-evaluation technique of the risk matrix, as depicted in Figure 5.8.

Figure 5.8 shows how the similar and the complementary features of the three techniques can be used to integrate them into an optimal combination to decide on the actions that can be taken to stop the development of external major incidents.

Part A of the figure includes a safety risk-identification performance on a chemical installation, for example, situated on the premises of a company *A* and within a certain distance to the boundary of a company *B*. This part of the process is an intra-company technical matter. In the example, Part A is done exclusively by personnel of company *A*.

The identification exercise is characterized by a bowtie structure. Central in the bowtie is the hazard, the left-hand side of the bowtie describes how events and circumstances, either in isolation or in combination, can release a hazard with the potential of harm to assets, people or the environment. The right-hand side represents the various scenarios that might develop from the undesired event.

In the first step of the hazwim process, thorough safety-risk identification determining all possible hazard causes and hazard consequences is performed using the hazop procedure. Hazop results typically include a list of recommendations that address the potential problems identified. Unfortunately, the results provide no guidance to decisions makers as to which potential problems are most

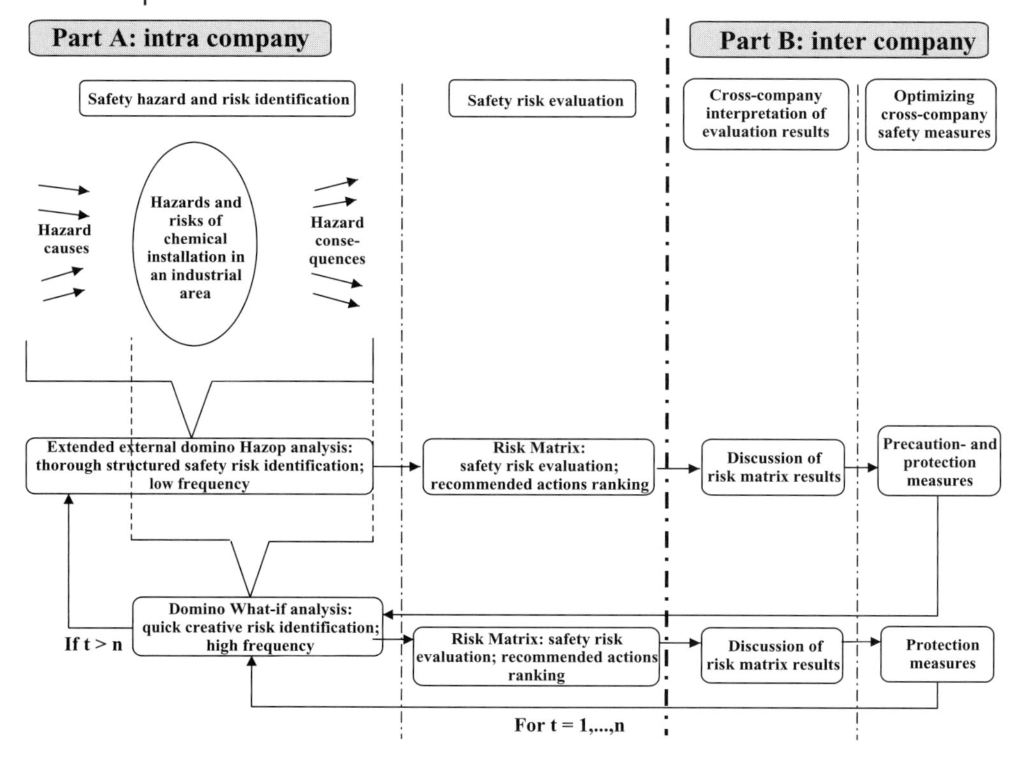

Figure 5.8 Hazwim process scheme.

important and which are of minimal concern. Therefore, the next step in the process is a safety-risk evaluation linked with recommended actions.

In Part B of the hazwim process, results of Part A are discussed by safety managers of both plants to decide upon which prevention measures to take. Part B is thus a cross-company management matter.

In the next step, performed on the same installation later in time, less expensive and less time consuming risk identification is executed using the what-if analysis technique. Here also the risk matrix is used to classify risks and to rank the proposed protection measures. This step, using a quick creative safety-risk-identification technique, can be performed several times during a certain amount of time to determine the installations' risk potential. The recommended frequency of what-if risk analysis performances, depicted in Figure 5.8 with "*n*", varies per installation depending on installation (danger) characteristics.

After *n* times investigating the safety risks of the installation performing a what-if analysis, a more thorough safety-risk identification by hazop analysis is executed next to ensure optimal external domino safety.

The hazwim process is thus a standardized but flexible scheme and time or personnel boundaries should be adapted depending on different factors and various circumstances.

Hazwim process implementations in a chemical industrial area are organized using the hazwim organization schedule as explained in Section 5.2.2 Step 4.

5.2.2.4 Discussion

Hazwim allows for performing a thorough cross-company external domino-risk analysis, but at the expense of a certain complexity level. The scheduling of the hazwim scheme sessions can be time consuming and demanding on the company's human resources. Highly qualified staff is indispensable and the requirements for the facilitator of the hazwim study are high: experience in hazop analysis, what-if analysis and the risk matrix.

On the other hand, by judging the cost effectiveness of performing hazop in combination with the what-if technique and by an optimization of applying the recommended actions as well as by preventing possible major accidents, the framework can be a very powerful optimizing tool to deal with external domino danger in a multi-plant area.

5.3 A Multi-Plant-*Security*-Management Framework

5.3.1 The Principles

Every chemical enterprise should implement a security-management system through identification of its value chain, for it is only by the identification of added value that a company enables to comprehend the need for security efforts (ES5, American Chemistry Council, 2006).

A security-management system has several vital functions that contain the actual general objectives of security. Careful implementation of the company's security-management system turns these functions into advantages that secure the survival of the chemical company. Accordingly, a security-management system can be interpreted as an economic investment.

First, a chemical enterprise's security-management system ensures protection of the plant's tangible and intangible assets (i.e. protection of all people, information and properties within the plant's premises). Secondly, it guarantees the business continuity of a company. This term expresses a set of processes and procedures being implemented during or after an incident or assault in order to assure the continuity of every essential process within the company. A security-management system assists in achieving this aim, in combination with safety- and quality-management systems (MI5, 2005).

In order to know *how* a chemical enterprise should be protected, the contents of a security-management system must be examined. Management practices depicted in a security-management system can be structurally divided into two main parts: (i) creating and maintaining a specific security culture, and (ii) development and implementation of security measures (ES6, American Chemistry Council, 2001).

Table 5.5 Eight-step security analysis method.

1. Value identification
2. Threat identification
3. Security-risk identification and evaluation
4. Identification of current security measures (physical and organizational)
5. Evaluation of existing security measures vis-à-vis existing security risks
6. Solution possibilities to prevent threats and/or to minimize their effects
7. Solution(s) proposition
8. Cost estimation

The existence of a company security culture is, therefore, very important for a chemical plant. A security culture implies all staff members sharing the same values, norms, behaviors and goals as regards security. Such a culture can only be developed and maintained if company top management takes all responsibilities required. To this end, the chemical enterprise needs to have a strong leadership and a senior management involvement, it also requires adequate resources being available to take security measures, which should be accompanied by a clear communication; and it should have the disposal of efficient and effective management of change procedures throughout the company.

The various security measures must be established and put into operation by using a specific security culture as a guide. The development takes place following a thorough security analysis, after which the actual physical and organizational security measures are implemented.

A security analysis should clearly demonstrate whether new security measures should be introduced, which old measures can best be replaced or whether no security-related changes are required within a company. In practice, a security analysis methodology at a chemical firm consists of eight consecutive steps, as illustrated in Table 5.5.

The different security measures are implemented after completing the above steps of the security-analysis method. An example of a security analysis methodology used in the US's chemical industry is the RAMCAP method (CCPS, 2003; Moore *et al.*, 2005 and 2007).

Physical security measures are generally characterized by three objectives: to deter, to detect and to delay all potential threats. The three security-risk-management strategies are explained in Table 5.6.

Based on the security-risk-management strategies of Table 5.6, a number of concrete measures can be defined. These measures can be personal, structural or electronic and they are present at different locations within the chemical company. Furthermore, the different locations on the plant site can be divided into three categories, that is, the perimeter,[5] the terrain and the diverse buildings and rooms. Table 5.7 further illustrates which physical security measure is related to what

5) The perimeter is the whole area between the site and the street and water's edge (in case a chemical enterprise situated in a port area does have a quay to receive ship suppliers) including the transition points between the actual perimeter and the site (such as, e.g., entrances and exits).

Table 5.6 Strategies to manage security risks.

Strategy	Definition
Deter	To prevent or discourage the occurrence of a breach of security by generating fear or doubt. Physical security systems, such as warning signs, lights, uniformed guards, cameras, are examples of systems that provide deterrence.
Detect	To identify an adversary attempting to commit a malicious act or other criminal activity in order to provide real-time observation, interception and post-incident analysis of the activities and identity of the adversary (e.g., intelligence serves this purpose).
Delay	To provide various barriers to slow the progress of an adversary in penetrating a site to prevent an attack or theft, or in leaving a restricted area to assist in apprehension and prevention of theft.

Source: based on CCPS (2003).

Table 5.7 Matrix of physical security components.

<table>
<tr><th></th><th>Perimeter</th><th>Site</th><th>Buildings and rooms</th></tr>
<tr><td rowspan="2">Personal</td><td colspan="3">Security guards</td></tr>
<tr><td>Reception</td><td colspan="2"></td></tr>
<tr><td rowspan="5">Structural</td><td>Fences</td><td colspan="2"></td></tr>
<tr><td>Entrances and exits[a)]</td><td></td><td>Entrances and exits</td></tr>
<tr><td colspan="2">Outdoor lighting</td><td></td></tr>
<tr><td colspan="3">Signalization</td></tr>
<tr><td colspan="2"></td><td>Windows and doors</td></tr>
<tr><td rowspan="4">Electronic</td><td>Electronic access control systems[b)]</td><td></td><td>Electronic access control systems</td></tr>
<tr><td colspan="3">Cameras (e.g., CCTV, outdoor cameras, etc.)</td></tr>
<tr><td colspan="3">Detection systems (e.g., sensors, alarm-monitoring capabilities, etc.)</td></tr>
<tr><td colspan="2"></td><td>Control room</td></tr>
</table>

a) Physical security measures at entrances and exits are successively gates, barriers and additional security devices (such as, e.g., rising bollards), tourniquets and the key plan.

b) The electronic access control systems include voice control systems, electronic badge systems and biometric access control systems.

location. The advantage of this model is that an overlap of physical security measures between two or more places becomes visible making it easy to use in practical circumstances.

It should, however, be noted that these physical security components are actually not unique to the chemical industry: the same security principles and security measures used by the chemical industry can be seen in certain other facilities as well, such as nuclear power plants.

According to the professional literature (Fennelly, 2004; Landoll, 2006; Fisher *et al.*, 2008), there are seven management practices or organizational security measures for guaranteeing a reliable plant-security-management system. These management practices should not be considered separately, as interdependencies between them are essential. Important security management practices can be summed up as:

5.3.1.1 General Security Policy

A chemical enterprise must have a clear formulation of vision, mission, strategy and objectives as regards security.

5.3.1.2 Organization, Planning and Documentation

For effective implementation of the general security policy, a formalized security organization, planning and documentation should be created. The organizational aspect is reflected in the security procedures of the chemical company, while the planning and the documentation are reflected in the existing plans and documents.

5.3.1.3 Communication and Cooperation

A chemical enterprise should be able to guarantee efficient and effective internal and external communication and cooperation. Internal communication and cooperation take place through top-down, horizontal and bottom-up interactions. External communication and cooperation should be guaranteed at local, national and international level.

5.3.1.4 Training, Education and Guidance

The competences of every security staff member of the chemical company should be kept up-to-date or should be increased by means of training sessions and educative courses. Besides these efforts, guidance for all employees is necessary in order to maintain security awareness.

5.3.1.5 Crisis Management

In case a threat is detected or in case of an actual terrorist attack, effective crisis management is essential. Following an incident or a near-incident, further (internal or external) investigations are required.

5.3.1.6 Audits

A chemical enterprise should periodically conduct internal and external security audits in order to evaluate the existing security measures and to adjust them, if necessary.

5.3.1.7 Third-Party Verification

Chemical enterprises should invite third parties (such as insurance agents or members of the local or national authorities) with the aim of verifying whether the promised or compulsory security measures have been adequately implemented

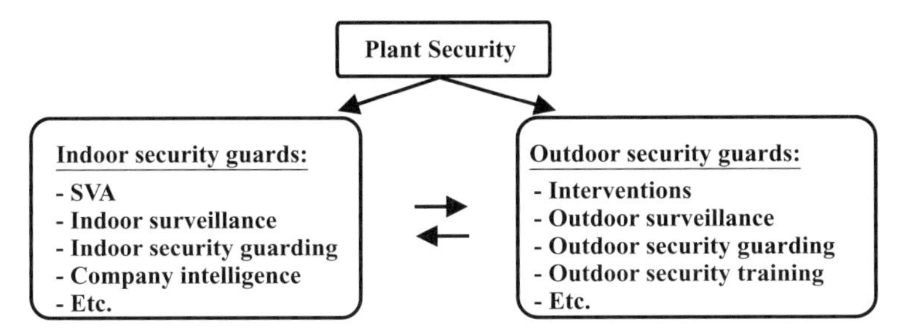

Figure 5.9 Indoor and outdoor security activities.

or modified. This verification ensures sufficient protection for the surroundings as well.

The question to be answered in this book is how to develop a framework that is able to facilitate the collaboration between security managers from different plants within a chemical cluster. Collaboration should be enhanced concerning physical as well as organizational security measures. This problem can be solved by looking upon security management from an "indoor" and "outdoor" perspective, a concept that is explained hereafter.

To date, no distinction has been made between indoor security and outdoor security. Nonetheless, routine work processes are usually more standardized and have more formalized coordination, while work processes requiring action in novel situations are less prescribed. Indoor security, including carrying out security-vulnerability assessments, is characterized by rather highly standardized procedures at company level, while outdoor security activities, including interventions, are usually less company specific. Figure 5.9 illustrates the difference between outdoor and indoor security tasks.

Designing adequate mechanisms of coordination between security personnel within and between organizations largely determines the vulnerability of these organizations. A key issue in this respect is the degree of standardization of such coordination. The Center for Chemical Process Safety (CCPS), American Society for Industrial Security (ASIS) and the American Chemistry Council (ACC), among others, have made an effort to increase the level of security standardization in the chemical industry to make chemical security more predictable and controllable.

Where indoor security is concerned, an important difference exists between techniques aimed at preventing malicious acts from within the company (e.g., by sleeper-cell employees[6] or by commercial spy employees), such as building up company intelligence, and techniques aimed at preventing attacks from outside the company (e.g., by terrorists), such as carrying out SVAs. Both kinds of indoor security activities are characterized by a high degree of standardization, but only the techniques for external attack precautions using information of a low confidentiality can be considered for collaboration with neighboring plants. A lower

6) "Sleeper-cell employees" indicate company personnel characterized by the potential, or waiting for the right moment, to "attack" one way or another their own employer.

standardization degree is beneficial for outdoor security tasks, since outdoor guards are less inclined to stay alert within highly harmonized settings. In this way the possibility that outdoor security guards lose the ability to switch to a highly flexible action mode when an unexpected threatening situation occurs, is reduced. Moreover, surveillance of the property fence line in a non-standardized way (e.g., at unpredictable moments) can prove to be a very effective security precaution measure. Of course, some outdoor preventive security tasks (e.g., gate control) remain highly standardized.

Summarizing, chemical corporations need to control diverging aspects of corporate security, including premises security, crime prevention, sleeper-cell employees' investigations, information security, personnel security, interventions, terrorism protection, *etc.* All of these aspects contribute to total plant security. The operating range of security management thus being wide, especially in the case of large plants, it is recommended that the responsibilities, activities and procedures be divided between outdoor and indoor security-guard services.

5.3.2
Achieving Solid Security for a Chemical Industrial Area

The fundamental basis of security management can be expressed in a similar way to the *layers of protection* used in chemical process plants to illustrate safety barriers (see Figure 5.1). In the similar concept of concentric *rings of protection* (CCPS, 2003), the spatial relationship between the location of the target asset and the location of the physical countermeasures is used as a guiding principle. Figure 5.10 exemplifies the rings of protection and their component countermeasures, illustrating the responsibilities and the power of indoor and outdoor security guards.

In security terms, critical infrastructure is broadly defined as people (employees, visitors, contractors, nearby members of the community, *etc.*), information (formulae, prices, processes, substances, passwords, *etc.*), and property (buildings, vehicles, production equipment, storage tanks and process vessels, control systems, raw materials, finished products, hazardous materials, natural-gas lines, rail lines, personal possessions, *etc.*) that is believed crucial to prevent major business disruption and resulting substantial economic and/or societal damage.

Figure 5.11 looks at the rings of protection from a different angle, using the idea that indoor and outdoor security management are separate services with their own responsibilities and activities. In the figure, indoor security is in turn divided into a section concerned with possible attacks from within the company (which is called indoor security against insiders or indoor security [I]), and a section responsible for preventing external attacks (which is called indoor security against outsiders or indoor security [O]). Making this particular distinction allows for the elaboration of frameworks where indoor security personnel are able to cooperate across organizations without having to discuss highly confidential data.

By considering the sequence of events that might lead to a potentially successful attack, another representation can be given, illustrating the effectiveness of the rings of protection (see Figure 5.12).

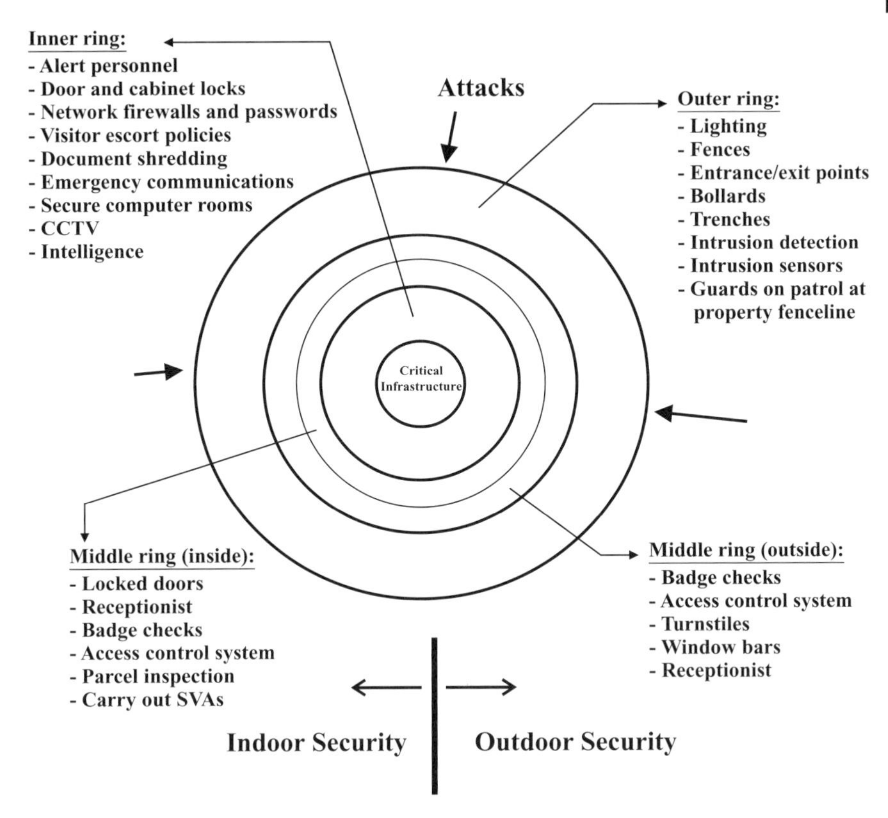

Figure 5.10 Security rings of protection (Source: based on CCPS, 2003).

First, companies can clearly protect themselves in a much better way against external attacks than against attacks from within the company itself, since in the latter case there only exists indoor security against insiders to avert the threat. Secondly, since the effective prevention, protection and mitigation of attacks depend on meticulously carrying out SVAs the latter is of crucial importance to deter, detect and delay possible threats within a single company as well as within a cluster of companies. Thirdly, efficiently protecting against external aggression implies actual collaboration strategies between cross-plant outdoor security services. The next sections discuss frameworks to encourage the elaboration and streamlining of cross-corporate indoor and outdoor security teamwork.

5.3.3
Shaping a Framework for Indoor Security Cooperation: InSec

One of the most important aspects in the overall security protection of multi-plant areas is carrying out security-vulnerability assessments. This vital task is the responsibility of indoor security against outsiders. The SVA process is a procedure

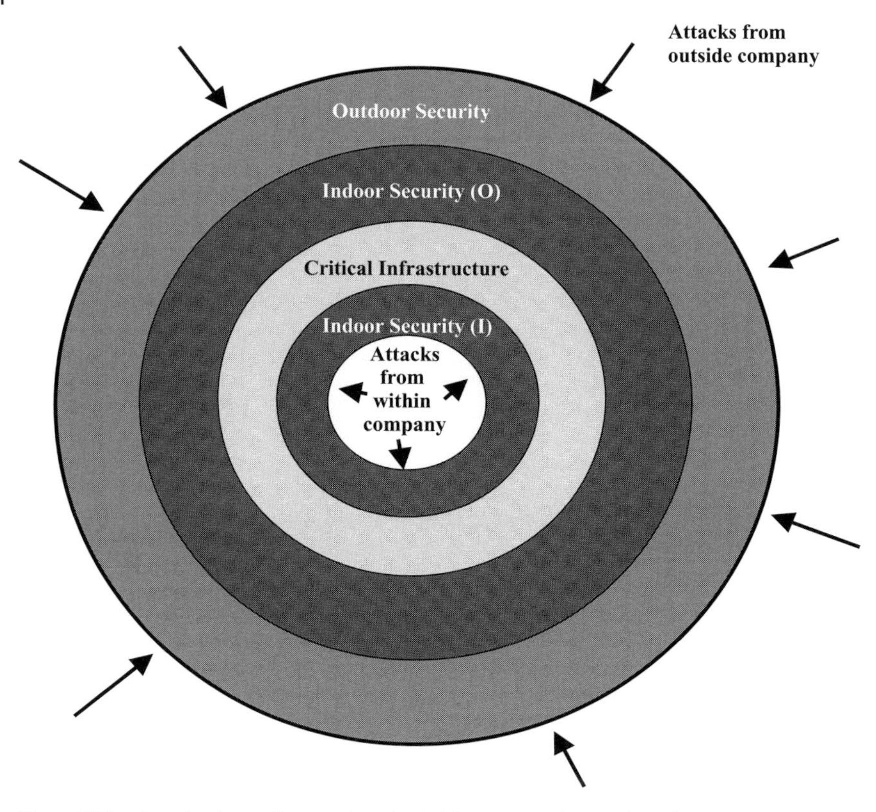

Figure 5.11 Security rings of protection divided between indoor and outdoor security management.

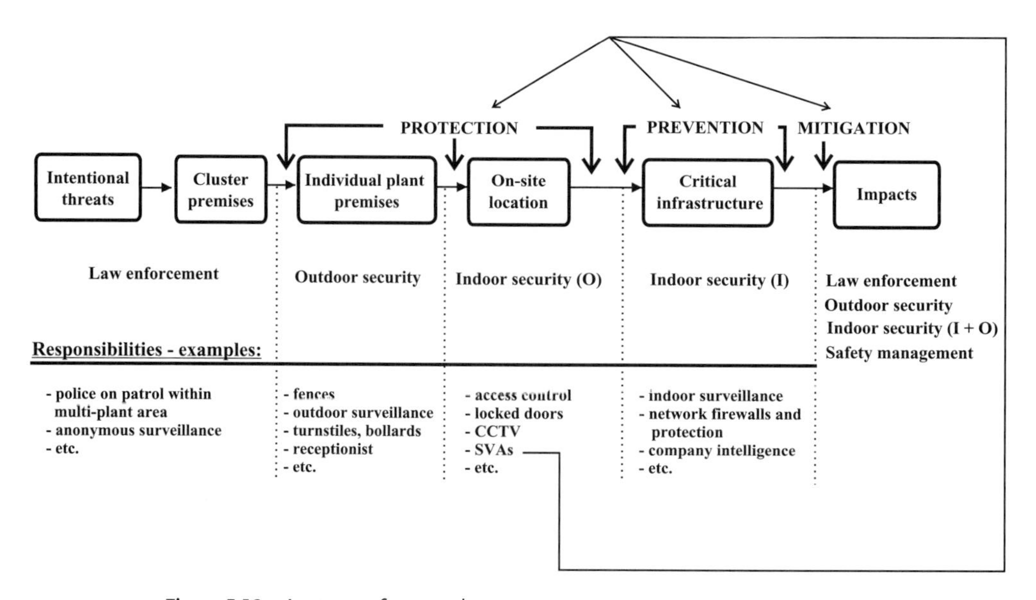

Figure 5.12 Anatomy of an attack.

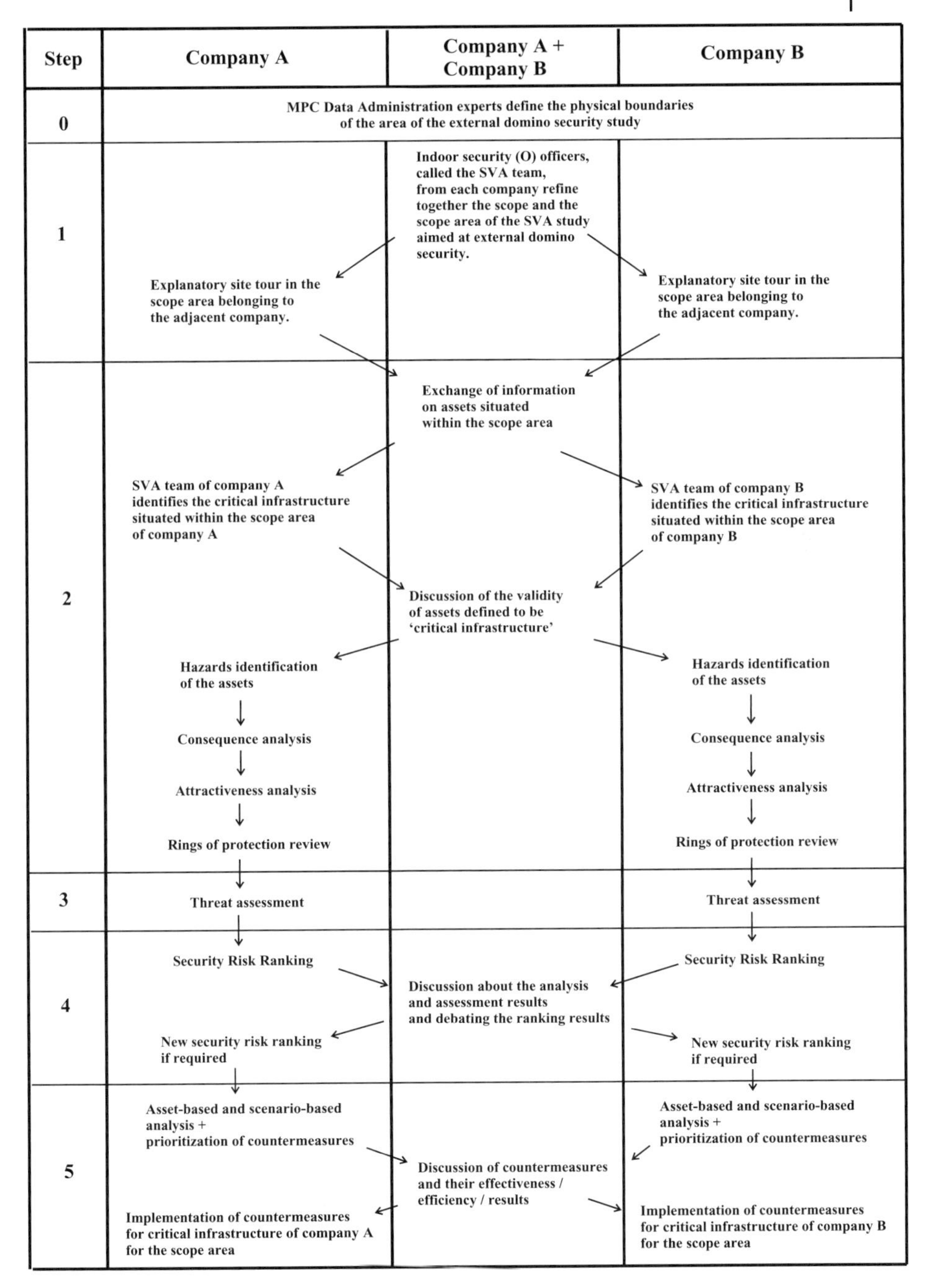

Figure 5.13 Framework "*InSec*" for enhancing cooperation between indoor security against outsiders services from neighboring companies.

that is similar to a safety-risk analysis, but that addresses security risks. It is the technique of determining the likelihood of an adversary successfully exploiting vulnerability, and the resulting degree of damage or impact. SVAs are performed qualitatively using the best judgment of security professionals. Estimating security-related risks is the desired outcome of the procedure, providing the basis for their rank ordering and thus establishing priorities for the application of countermeasures.

For an SVA to be thoroughly carried out by competent indoor security against outsiders personnel every security risk pertaining to an organization should be covered. However, security risks, like safety risks, are not contained within the borders or the fences of an enterprise. External domino accidents can be induced by an adversary. Therefore, to accomplish the identification of potential cross-company malicious acts, indoor security against outsiders of neighboring companies needs to undertake joint SVAs. The SVA procedure involves five successive implementation steps: (i) project planning; (ii) critical asset characterization, (iii) threat assessment; (iv) vulnerability analysis; and (v) identification of countermeasures. Each of these steps includes several successive sub-steps. Since confidential information is often used in the course of carrying out the different steps and sub-steps, companies are bound to be reluctant to perform these stages jointly. Nevertheless, intentional domino risks with the potential to affect adjacent infrastructure should be security protected. To overcome the potential security weakness of external escalation risks through means of collaboration, a framework called InSec is proposed, illustrated in Figure 5.13.

In the preliminary step, multi-plant council data administration experts define the physical boundaries of the domino-security investigation. Using a newly developed decision-support tool to be used within the multi-plant area for dealing with external domino-effects prevention, Chapter 6 discusses the division of the multi-plant area into sub-areas (called domino islands) to decrease the possibility of domino events. The stage afterwards (i.e. the first step), the SVA planning step, is important for the remainder of the cross-border SVA study. Security officers from the different companies participating in the SVA exercise refine the scope of the study, that is, the assets (people, chemicals used, chemicals produced, environment, equipment, activities, *etc.*) that need to be investigated per domino island. Afterwards, an explanatory site tour for indoor security against outsiders experiencing all adjacent plant's assets belonging to the scope area is organized. A reconnaissance tour such as this will contribute to the indoor security against outsiders's knowledge to be used at a later stage and to the mutual understanding between the teams.

The second stage, in which critical assets are to be characterized, starts with an exchange of information between SVA teams concerning all assets that might pose a danger to the neighboring company (e.g., risk contours of infrastructure situated within the scope area). Using this information as well as a company's own data, SVA teams develop a potential target list of assets that require further attention regarding cross-company threats. This ranking is discussed by joint SVA teams eventually leading to a definitive list. The latter inventory is used by the individual teams as a listing of assets of consideration. The hazards need to be identified, the consequences of a successful attack on the assets need to be assessed, as do the attractiveness of a target, and the existing rings of protection need to be identified and documented for every asset listed.

Based on the activities performed in the previous step, the threats to the identified targets are characterized. As regards the security-vulnerability analysis, this means developing model adversaries in terms of their characteristics and capabilities.

In the fourth stage, the actual vulnerability analysis is accomplished. Using an asset-based as well as a scenario-based approach to vulnerability analysis (see CCPS, 2003), cross-company security risks are ranked. In other words, the level of risk of the adversary exploiting the asset given the existing security countermeasures is determined. To this end, a user-friendly harmonized instrument familiar to all participants can be used, which is a special version of the risk matrix, that is, the *security-risk matrix*. Figure 5.14 proposes such a security-risk matrix. The security-risk matrix results are discussed by the neighboring indoor security against outsiders' teams, eventually leading to adapted plant security risk rankings.

In Figure 5.14, an example is given of an asset characterized by prioritization criteria with a total of 18 (8 + 3 + 7). The latter numerical value should be viewed relative to other total prioritization values of assets.

The last step involves the planning of countermeasures by both companies. For the selection of countermeasures, the two assessment approaches are also used, that is, the asset-based and the consequence-based approach. Afterwards, the effectiveness of deterring or detecting an attack and of delaying an attacker until appropriate authorities can intervene, is discussed between all participating SVA teams. The last sub-step involves single-plant indoor security against outsiders implementing the countermeasures for the scope area as previously discussed and approved in their own plant.

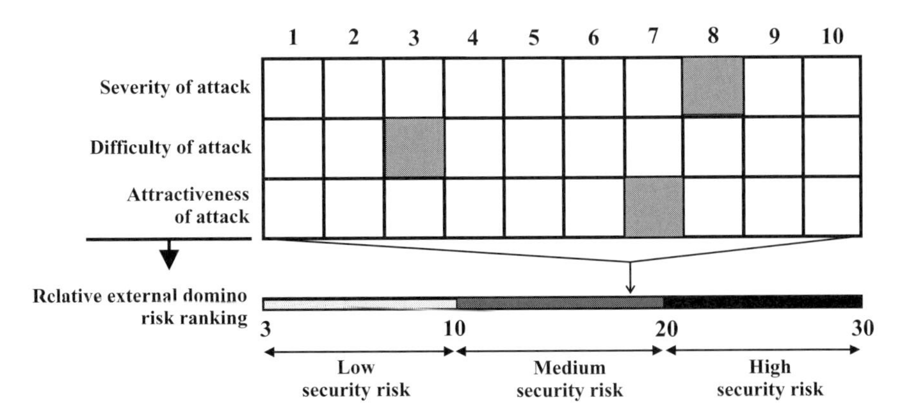

Figure 5.14 Exemplary filled in security-risk matrix for a specific asset.

5.3.4
Shaping a Network for Outdoor Security Cooperation: OutSec

Outdoor security management cooperation depends on the existence of mutual trust between outdoor security teams belonging to different plants and the establishment of well-considered structures and responsibilities. Outdoor security teams can in fact be bonded relatively easily by the continuous "strife" against a common adversary external to the chemical multi-plant area. Solid inter-team relationships need to be formed to create an efficient and highly effective network of outdoor security teams. To achieve this objective, an intensive multi-company training program for all outdoor security teams is required. Minimum training criteria from a holistic multi-plant initiative perspective need to be established. Training may consist of field training, task-oriented training, computer-based training, classroom training, self-study, psychological training, *etc.* Training topics include observation and incident reporting, principles of communications, principles of access control, principles of safeguarding information, emergency response procedures, life safety awareness, basic knowledge of chemical substances belonging to the team's area of surveillance, crisis management, conflict resolution awareness, crowd control, *etc.* It is further recommended that all training is accompanied by appropriate assessment and evaluation to measure training efficiency. The objective is to create coherent cluster outdoor security teams that form a network and have an overview of each other's competencies. A competence network as illustrated in Figure 5.15 is shaped.

In such an outdoor security network, security teams do not merely use their own competencies and experiences (i.e. internal plant competencies and shared single plant competencies) but also rely on expertise and skill from external company Outdoor security teams (i.e. shared multi-plant competencies and external plant competencies). By shaping such a framework of expertise and skill through the use of well-elaborated and intensive single- and multiple-plant training programs, a hybrid outdoor security organization is developed. The latter structure guarantees the desire for, and in fact the requisition for, outdoor security services to be efficient, effective and highly flexible at all times and under any circumstances.

Furthermore, to guarantee adequate security, it is crucial that the manning levels of indoor security as well as outdoor security (quantitative as well as qualitative) are sufficient. An approach to assess security staffing levels is proposed in the next section.

5.3.5
Assessing and Evaluating Security Staffing Levels in a Multi-Plant Area

Drafting an evaluation instrument that explicitly addresses indoor and outdoor security staffing levels is highly relevant because of the significant role played by the human factor in plant and multi-plant security. The greatest dangers and threats concerning chemical industrial area security stem not so much from easily

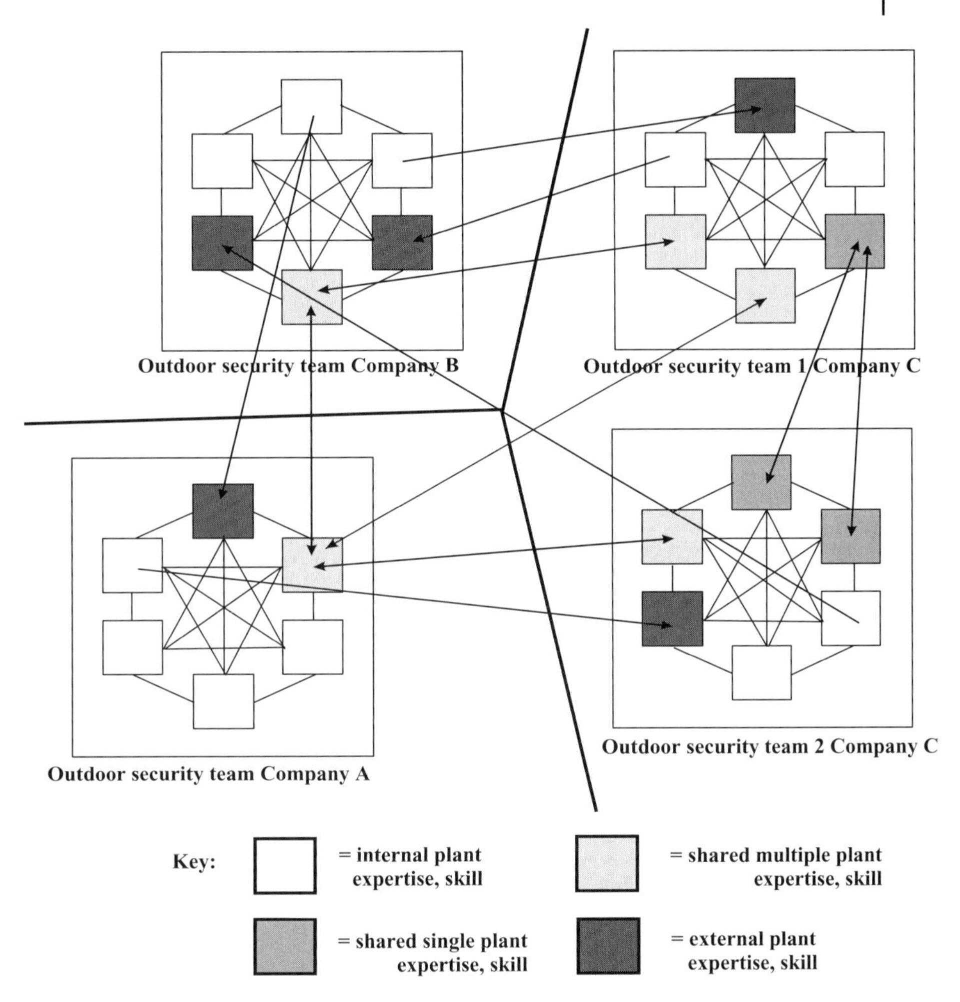

Figure 5.15 Network of outdoor security guard competencies.

detectable mischief as from thoroughly planned and highly covert malicious acts. Actions such as these that are difficult to observe will only be noticed if the quality and quantity of indoor and outdoor security staffing is sufficient. Therefore, base level organization of security manning, including security personnel task allocation, definitions of roles and responsibilities, and surveillance practices, is very important. Good security staffing design includes more than the mere assurance that cameras are located on the premises of an industrial area. Critical infrastructure needs to be identified and monitored 24/7, and security personnel might need to be present within minutes to be able to prevent intentional catastrophes.

Establishing a document to judge whether security staffing is sufficient remains a far from easy task. A methodology for evaluating security manning is worked out and is explained hereafter.

Different possible physical configurations within the indoor and outdoor security manning arrangements exist:

a) Permanently present indoor and outdoor security guards are responsible for surveillance of the whole multi-plant area;

b) Outdoor security guards, not permanently present at the site but who may be summoned from elsewhere within the plant or multi-plant area, are responsible for outdoor interventions within the entire multi-plant area;

c) The industrial area is divided into different sub-areas in which distinct outdoor security personnel is responsible for outdoor surveillance and interventions (being permanently present or not);

d) Indoor security guards are permanently present for indoor and outdoor surveillance tasks;

e) A part or whole of the surveillance task is carried out by contracted security guards;

f) Mixed situations of the above possibilities.

If security staff were unable to be present in a particular location in the industrial area within a given time period, serious mischief might result. Different outcome scenarios have to be identified and the reliability of the surveillance equipment and the knowledge of the guards about how to intervene when required are questioned. Therefore, conditions such as the number of guards necessary, and the means and the competences needed to be able to secure the area are checked against the six principles given in Table 5.8 (Reniers *et al.*, 2007a).

These principles are questioned by checking whether (i) the principle can be violated; (ii) in which way the principle has been or could be violated; (iii) which measures are required to counter the violation; and (iv) whether these countermeasures are reliable and effective. In this way, the evaluation document identifies potential organizational bottle-necks of security staffing and indicates solutions for threatening situations.

The instrument, which amounts to a checklist that is provided as Appendix A, puts forward a number of binary (yes/no) questions related to the six principles of Table 5.8. All questions combined give an indication of the effectiveness of protecting the critical assets of the area. The checklist answers should provide a clear insight into the ability of an industrial area to handle indoor and outdoor security problems (especially problems related to more than a single company). As a reminder for the security officer, control questions (indicated "C.Q.") are inserted for each question, giving advice about what evidence is required to answer the question in the most unbiased way possible. This C.Q. is an open question asking for evidence, information and documentation to support the "yes" or "no" rating on the binary question.

Secrecy is a foremost requirement when it comes to security information of a chemical industrial surrounding. Clearly the questions contained in the security

Table 5.8 Parameters and principles to deal with security staffing level (SL) problems.

Security SL problem parameter	Principle	Interpretation
Security SL Problem Detection	1. Supervision / intervention possibility	There should be continuous supervision of the identified critical vulnerable assets by skilled security guards and the ability to intervene whenever needed.
	2. Distractions	Distractions such as answering the phone, talking to people, performing administrative tasks and acting upon nuisance alarms should be minimized to reduce the possibility of missing/overseeing/responding too late to security alarms and threatening situations.
Security SL Problem Diagnosis	3. Information	Security staff should have at its disposal sufficient information as well as the competence required for diagnosis of the problem and evaluation and handling of the situation.
	4. Communication links	Communication links at single plant level as well as at multi-plant level (in)between the (private or company) security-guard departments and law-enforcement agencies should be reliable to intervene whenever necessary.
Security SL Problem Handling	5. Assisting personnel	Sufficient security staff required to help in the case of highest threat level situations should be able to be present at every location in the area within a pre-defined time period.
	6. Handling Operations	Security staff should be allowed to concentrate on solving the threatening situation. Necessary but time-consuming activities should therefore be allocated to others, for example, summoning law-enforcement agencies or communicating with adjacent plant-security management. Moreover, all threatening situations should be solved as fast as possible.

staffing evaluation instrument cannot be posed to security guards or plant operators or personnel at random. Instead, the questions should be revealed to and subsequently investigated by a multi-plant or plant security officer. The security officer has at his/her disposal all documents he/she believes are necessary to perform the checklist. He/she is able to request that experiments (e.g., simulations, reliability studies, *etc.*) be carried out whenever needed.

Three parameters are used for evaluating security staffing arrangements, that is, detection, diagnosis, and handling of security problems. For each of these three issues, two principles are checked in the evaluation instrument. The evaluation of every principle, and hence every question, leads to a security staffing ranking

Table 5.9 Evaluation tabulation to be filled in.

Problem Parameter	Ranking	A	B	C	D	Total ranking
Detection	Principle 1	1A	1B	1C	1D	
	Principle 2	2A	2B	2C	2D	
Diagnosis	Principle 3	3A	3B	3C	3D	
	Principle 4	4A	4B	4C	4D	
Recovery	Principle 5	5A	5B	5C	5D	
	Principle 6	6A	6B	6C	6D	

Table 5.10 Description of the different grading for the security-staffing-evaluation instrument.

Degree	Description
A	The parameter of the security problems is guaranteed by the inherent presence of a sufficient quality and quantity of security staffing levels. The organization of the industrial area personnel guarantees security in all circumstances.
B	The parameter of the problem is guaranteed. There is no need for any kind of back-up system to solve the security problem. However, there is one disadvantage: the quality and/or the quantity of the staffing levels are not sufficient to guarantee the inherent presence of competent security personnel, information and/or communication in the case of some high-threat circumstances. The problem can be tackled promptly addressing the qualitative and/or quantitative security-staffing levels required.
C	The quality and/or the quantity of the security staffing levels suffice to solve threatening situations thanks to the presence of back-up systems. Measures are needed to ameliorate the response rate with which the security staffing levels are recovered and/or to ameliorate the quality and/or the quantity of the security staffing levels.
D	The organization of security personnel fails. For this problem parameter, security staffing levels should ameliorate in a qualitative and/or quantitative manner. The industrial area is not capable of guaranteeing security and preventing intentional incidents in the case of (some) highly threatening situations.

ranging from "A" to "D". Table 5.9 offers an easy-to-use ranking tabulation, indicating the company security staffing levels' failures.

In the "total ranking" boxes, the worst of the combined ranking outcomes is always assigned for each problem parameter. For example, if ranking principle 1 = 1A, and ranking principle 2 = 2C, then the total ranking for detection of problems would be "C". The implications of the ranking degrees A, B, C or D in terms of the final decision are given in Table 5.10.

5.4 Summary and Conclusions

The use of inherent safety is generally accepted to be the best long-term approach to chemical process safety. Inherent safety process improvements can be described as those that essentially eliminate hazards from the process. Inherent safety is obviously safer than controlling the hazards with even the safest or highest integrity safety instrumented functions.

Thus, using the appropriate techniques aimed at inherently safe process design during the process of constructing a new chemical installation, helps assuring long-term effective and efficient implementation of safety measures and leads to cost-effective facilities. The importance of ensuring that independent, diverse protection layers meet the specifications of the process hazards analysis team cannot be stressed enough. It guarantees the best possible total safety. Only when the design intent is met and maintained, is the total risk control strategy for the plant effective. Implementing process hazard analyses developed for enclosing inherent safety into the process design is the first requirement during the design phase.

The majority of Seveso top-tier companies have expressed a willingness to cooperate more intensively to protect themselves against potential off-site major accidents. The synergetic combination of the two complementary risk-analysis techniques of hazop and what-if analysis offers cooperation opportunities for all the participating companies to tackle external domino risks. Three techniques, hazop, what-if and the risk matrix, actually appear to be three of the four most commonly used techniques within the chemical industry. The benefits of all three approaches are their relative simplicity and the requirement of limited resources. This makes them very popular and thus the associated know-how and expertise is widespread and easily available within chemical companies. They can be effectively used to qualitatively or semi-qualitatively analyze and evaluate external domino risks. The popularity of the three techniques is crucial for a proposed safety collaboration framework, hazwim, to become a solid standardized method to be widely used in the chemical industry. The proposed meta-technical hazwim process appears to be promising in a number of respects. It combines safety-risk analysis with safety-risk evaluation, the two fundamentals of good safety-management practice. Moreover, the process combines individual company policy with cross-company insights to decide upon precaution and prevention measures for hampering external domino effects.

The hazwim framework can be used in a given setting of installations. Using the techniques of hazop analysis, what-if analysis and the risk matrix, plant-safety managers communicate at the same level to take cross-plant risk decisions (e.g., the taking of preventive measures). Hazwim recommends how to avoid potential trouble-inducing situations such as the swapping of confidential information that could cause difficulties for mutual collaborator trust. The structure of the hazwim framework does not limit the creative and imaginative power of the hazop or what-if teams. External industry evaluation suggests that the hazwim framework, supported by the hazwim process scheme and the hazwim organization schedule, is a very useful instrument for company external domino safety policy decisions. It tries to offer a priority list of recommended actions to eliminate or alleviate the most dangerous off-site hazards in an effective and efficient way.

Furthermore, the effective management of security issues is an ever more fundamental requirement of chemical business. Therefore, a framework called InSec is suggested for enhancing cross-plant security-vulnerability analysis cooperation and leading to countermeasures and rings of protection covering not only single-plant asset vulnerabilities, but also multiple-plant vulnerabilities (such as those caused by external knock-on security risks).

As a last contribution to engineering multi-plant management frameworks, a hybrid network concept is suggested for competent outdoor security services to share competencies, expertise, training sessions, operational knowledge and accessibility, among others. Experience, training sessions, know-how and knowledge are divided across the different teams, allowing for them to intervene adequately and to be extremely flexible in extremely exceptional circumstances, such as terrorist attacks.

The quality and the quantity of security staffing levels required to perform surveillance tasks and intervention activities is basic in an efficient single-plant and/or multi-plant security policy. This chapter (and Appendix A) provides a user-friendly checklist for the security officer to evaluate security manning levels in an industrial area. The evaluation instrument provides guidance on how to meet the requirements of adequate quantitative and skilled security personnel.

6
A Multi-Plant Safety and Security Culture – The Technology: Developing the Tools to Advance Multi-Plant Safety and Security

6.1
Introduction

Solid major accident-prevention management is characterized by efficient and effective risk assessments. As a means of addressing the efficiency aspect, decision-support-analysis software is becoming increasingly available. Nonetheless, there are relatively few software packages available to study (internal or external) primary domino accidents in complex industrial areas and even fewer to forecast potential catastrophes caused by secondary order (involving a sequence of three installations submitted to two consecutive accidents), tertiary order or even higher-order accidents. Moreover, available domino software usually focuses on risk assessment and on consequence assessment. To the best of the author's knowledge, none of the available toolkits specifically address the prioritization of installation sequences in an industrial area composed of a number of chemical plants in order to facilitate an objective ranking of installations to determine where to carry out risk analyses concerning external domino effects.

Dividing a chemical industrial area into smaller sub-areas between which no domino effects are possible (regardless of installation failure probabilities), is a possible technique to decrease domino-security-risk impacts. A more thorough examination of these sub-areas for installations having a high contribution to possible domino effects (including installation failure probabilities) is proposed as a method of preventing unwanted domino-safety-risk outcomes.

Optimizing the decision process for taking prevention measures in a complex chemical industrial surrounding requires quantifying the danger (with respect to domino effects) of the entire industrial area, creating sub-areas, succeeded by domino danger quantification of every possible path giving rise to domino effects in these sub-areas. Available data should allow for ranking the domino paths in the installation networks and sub-networks and allow for objective safety and security management decisions. The computer-automated decision-support tool to this end should be developed by the MPC of the multi-plant initiative. This book provides the basic ideas and approaches that can be employed for developing such required multi-plant technology and operational tool(s). Using a holistic approach and thus looking at the multi-plant area as a whole, the tool should

Multi-Plant Safety and Security Management in the Chemical and Process Industries. G.L.L. Reniers

ISBN: 978-3-527-32551-1

allow preventive measures that minimize the impacts from potential intentional attacks and from potential accidental escalating accidents within the area to be taken.

6.2
A Multi-Plant Domino-Risk Methodology and -Decision Support Tool

6.2.1
Prevention Optimization in Industrial Areas: A Theoretical Domino-Effects-Evaluation Model for Developing Domino-Risk Software

6.2.1.1 Drawbacks of Current Domino-Risk Software in a Multi-Plant Context

Although most available software provides its user with application guidelines, offering data sheets, selection flowcharts, *etc.* the amount of data needed is quite extensive and requires considerable expertise of the person or persons using the tool. A substantial amount of time is required to gather and process the data. This is a first important drawback of current domino-risk software. Moreover, some of the data required for analyzing external domino risks should be provided by different plant safety and/or security managers from nearby companies, possibly resulting in confidentiality concerns and difficulties. No software framework is proposed to deal with these potential problems.

A second drawback involves the very nature of domino effects, looked upon from a purely deterministic viewpoint. A deterministic viewpoint upon domino software is ever more important from a security point of view, since it is impossible to determine the probability that a group of terrorists will simultaneously attack a certain number of installations within a chemical cluster.

Consider an installation *j* as illustrated in Figure 6.1, representing the highest danger for its surroundings according to the software results, that is, *j* has the highest dangerousness indication. Based on these results, the software user can

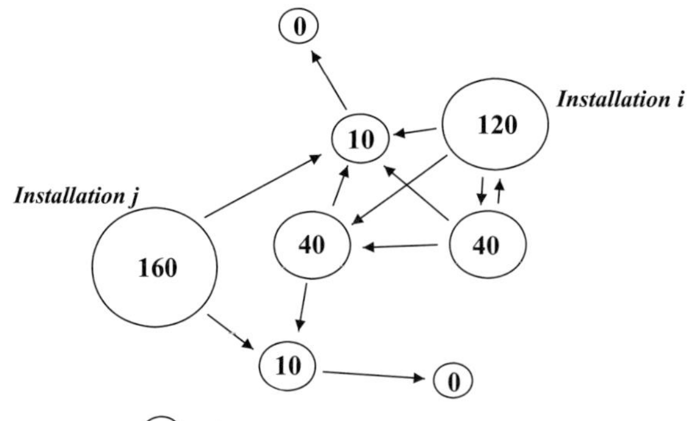

Figure 6.1 Fictitious example of an installation's network.

conclude that j needs – more than the other installations in the neighborhood – a preventive external domino-risk analysis approach. However, taking such pro-active actions as a consequence of these software results can be a sub-optimal approach to preventing domino effects in the area. Moreover, it can even prove an approach that leads to potential danger if the tool does not take into account domino effects involving more than two installations. Most available domino software tools at present suffer this potential drawback. The myopic (deterministic) model that would be currently used only considers "causing" and "receiving" installations, and not looking at 3-, 4- or even 5- installation sequences. However, the high severity of domino-effect consequences for installation j does not imply by default that this installation is the most significant domino-chain node in a complex installations network with respect to taking precautions in the network.

In Figure 6.1, a possible indicator for the danger in terms of domino effects is suggested for every installation within the circles representing the installations. Potential domino effects and their directions are illustrated with arrows. According to the used software it is apparent that installation j seems to be most dangerous in terms of domino effects. However, if more than 2 installations are taken into consideration in the domino accident calculation, it becomes clear that it is more efficient for the prevention manager to first take precautions against installation i.

In an analogous way, it is possible that following the result that an installation suffering the highest potential damage in the area according to the software tool, the software user would decide to protect this installation instead of protecting less vulnerable installations in the area (according to the software results); and by doing so neglecting the very nature of domino effects, that is the cascade potential.

Hence, those (majority of) software packages failing to include domino-chain accidents with three installations or more in deterministic domino studies, are not capable of determining in a pro-active and effective manner the most dangerous installation sequences in terms of escalation in a multi-plant area.

A third drawback of a lot of current domino-risk software concerns danger indicators and vulnerability indicators to be installation-specific values. However, the term "domino effect" actually implies having a sequence consisting of successive installation pairs. Therefore, it is more accurate to indicate the dangerousness of such "installation couple–arcs" instead of the dangerousness of single installations, that is nodes. Moreover, using dangerousness data linked with "the arrows in the network" allows for more objectively ranking installations sequences by making a distinction in dangerousness between the different sequences (see Figure 6.2).

The fourth drawback worth mentioning concerns the practical use of current available domino software tools. Irrespective of the availability of domino software packages on the market, their actual use in chemical enterprises is very limited and is restricted to drawing risk contours of potential internal and external domino risks. This observation brings about the possible problem of choosing an existing

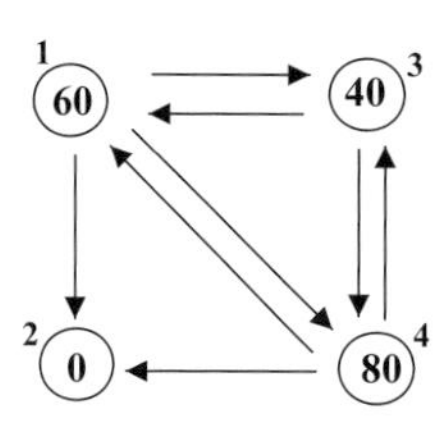

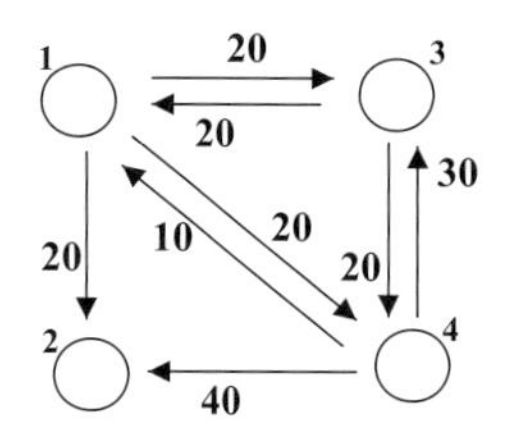

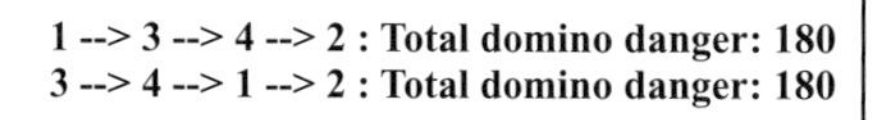

1 --> 3 --> 4 --> 2 : Total domino danger: 80
3 --> 4 --> 1 --> 2 : Total domino danger: 50

Figure 6.2 Distinction between total chain domino danger derived from single installation danger units and total chain domino danger derived from installation-couple danger units.

domino computer-automated tool in a multi-plant setting where the individual plants' cultures may considerably differ.

Therefore, and also because every multi-plant area has different characteristics and features and hence different needs, every MPC data administration should develop tailor-made software to tackle external domino effects in its specific multi-plant area.

6.2.1.2 Approach for Elaborating Industrial Area Knock-On Software

Building a user-friendly software tool for investigating knock-on effects related to safety risks as well as security risks requires viewing a chemical cluster in a holistic way. Domino-effect investigations should be carried out regardless of company borders and perimeters. Moreover, the software should be easy to understand and applicable to all companies involved.

By eliminating plant borders for the purpose of the exercise, an industrial region can be divided into a number of separate (chemical) installations. These installations can be considered as nodes of a network and the level of danger between the installations can be represented by the weight of the edges linking the nodes. Using this approach, a relatively simple network of connections or links emerges instead of a complex chemical industrial area.

Once the area is characterized by a single network of installations, it can be divided into smaller networks, that is smaller "clustered chemical installations", where there are no links between the sub-areas.[1] Such sub-areas are called *domino islands* in this book. The latter approach makes the region much less vulnerable to terrorist threats since the multi-plant area can no longer be destroyed in one go; only the sub-areas might break down entirely. Figure 6.3 illustrates

1) There exists no correlation whatsoever between installation-failure probabilities and terrorist attacks. Therefore, when splitting the multi-plant area into several sub-areas, no failure probabilities should be taken into account. Hence, only probability-independent consequences of domino-effect scenarios serve as guiding parameters to divide the multi-plant area into smaller areas.

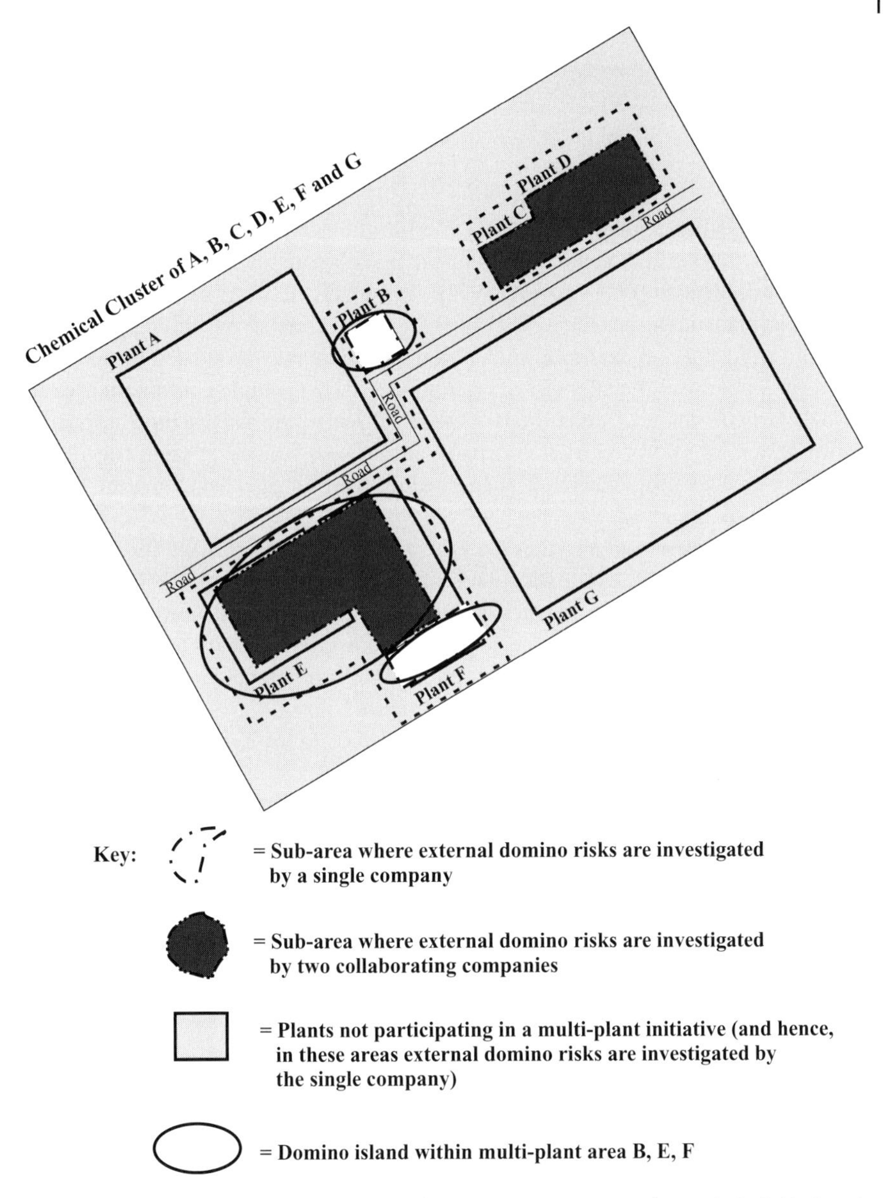

Figure 6.3 Example of an industrial multi-plant area comprising six plants (*A*, *B*, *C*, *D*, *E*, and *F*) divided into sub-areas.

dividing the multi-plant area *B*, *E*, *F* (originally vulnerable when broken down as a whole) into smaller areas leading to *domino islands* between which no domino effects can occur. Note that companies not forming part of the multi-plant area may be affected as well. Therefore, although plant *B* does not pose a threat to plants *E* and *F* (because companies *B*, *E* and *F* form one multi-plant area and plant *B* is recognized as a domino island within the multi-plant area), plant *B* for example has to investigate its external domino effects towards plant *A* and plant *G*.

To prevent the domino islands from collapsing, all different neighboring companies belonging to the sub-areas need to collaborate with each other, carrying out additional domino-risk analyses leading to additional domino precautions within the islands (as indicated in Figure 6.3). Within the sub-areas, the software should then allow determination of the chemical installations subject to the so-called design basis threats (DBT) (IAEA, 1998). These are the basic security threats to the multi-plant area, possibly causing man-made disaster by inducing external domino effects. As an example, such a domino effect may be caused by an attack on several chemical installations at once executed by multiple coordinated teams or by a large number of individuals. Possible threats may also include assistance in an attack by one or several company employees, suicide attacks, the use of explosive devices, *etc.* Protection against multi-plant design basis threats should be ensured by the MPC.

The new software tool should also take accidental domino risks into account. To meet this objective, the installations of every sub-area should be investigated for their propagation characteristics in the area (including installation-failure probabilities and scenario consequences). The top-ranked installations of this examination are rendered very safe, implying the most significant installations of continuing accidental domino events to have been eliminated.

Figure 6.4 illustrates the overall concept of making an industrial region safer and more secure as regards external domino effects.

Current available rather complex-to-use tools generally require extensive knowledge and expertise on the part of the user, such as, for example, an understanding of transient interchange of damage load (fire and explosion) between the causing installation and the receiving installation and mechanics of vessel failure. The available toolkits obviously have some very important advantages, allowing for a very thorough understanding of highly complex scenarios and generating all possible consequences that could occur at the installation investigated. Often, easy-to-read results in a graphical format are available as well.

However, both the complexity of the data required for assessing escalation events and the level of detail of the proper toolkits can prove a disadvantage for managers who have to decide about their use in a company. In chemical enterprises, safety managers are mostly interested in easy-to-handle and user-friendly decision-support tools, providing them with straightforward information ready for implementation.

Therefore, a theoretical conceptualization of how to manage – in a relatively simple way – a potentially very complex industrial cluster concerning the prevention and the mitigation of (internal and) external domino effects is described hereafter. The theoretical idea behind treating different chemical companies as one entire industrial multi-plant area is explained and the feasibility of integrating *low-probability, high-consequence* safety and security matters into a software tool is examined.

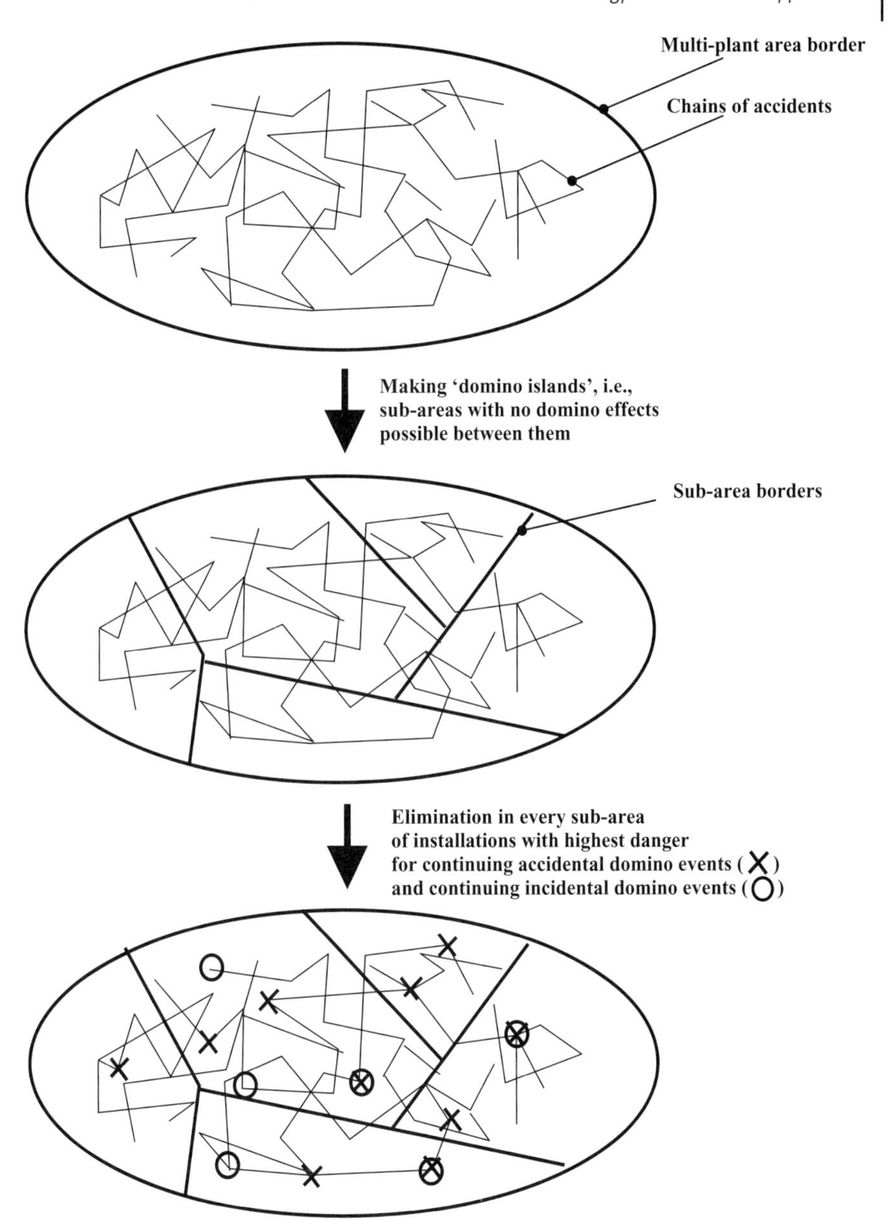

Figure 6.4 Conceptualization of domino software objectives for the improved protection of a chemical industrial area against external domino-safety risks and external domino-security risks.

6.2.2
User-Friendly Software for Planning Domino-Effects Prevention in a Multi-Plant Context

6.2.2.1 Introduction

Present safety research has led to several methodologies that assess the significance of safety-risk-related domino effects for major hazard sites. Factors relevant to domino escalation and various direct and indirect mechanisms that can produce a domino accident (caused by domino effects) have been determined. However, Atkins (1998) shows that an overall coherent approach for prevention optimization of cascade accident risks in a chemical industrial complex, does not exist. Recent empirical research in the Antwerp port area (Reniers and Soudan, 2003) indicates that this situation has not changed since 1998. Moreover, no *widely accepted* methodologies are available in literature for the assessment of domino hazards (Cozzani *et al.*, 2005).

Domino-safety research mainly focuses on identifying potential domino effects and determining their overall impact on the accident area. Often, mathematical models for the simulation of domino accidents are demanding due to the complexity of the accident evolution (simulation of the source term, direct ignition, non-ignition, delayed ignition, dense vapor, buoyant vapor, *etc.*) and to the complexity of the input data required. Nevertheless, several software packages have been developed that identify potential domino effects. Survey results (Reniers *et al.*, 2006) mention nine toolkits covering internal or external domino risks or the combination of both: AIDRAM-CARGO, CHARM, DOMINOXL, ExTool, FLACS, Shell Shepherd, ORBIT, SAVE and THESIS. Another toolkit developed with the specific aim of studying domino effects discussed in the literature is DOMMIFECT (Khan and Abbasi, 1998a) which is based on a systematic domino method, "domino-effect analysis" (DEA), also developed by Khan and Abbasi (1998b). The former (DOMIno eFFECT) is a software tool developed for domino-effect analysis in chemical process industries and is based on deterministic models used in combination with probabilistic analysis. Research of the DEA authors indicate, among other things, that it is not necessarily the installation of a plant that may potentially cause the biggest stand-alone accident that will also prove to be the installation most likely to cause a domino effect (Khan and Abbasi, 2001).

None of the existing methods is able to give information on the relative unidirectional danger between two installations in a chemical surrounding and by determining a domino-accident network, identifying the most dangerous domino-inducing installations in the area. To determine whether an unwanted event occurring at installation v_i is likely to give rise to an unwanted event at installation v_j, it is necessary to consider the magnitude of the occurrence at installation v_i, the likelihood (if the event is accidental) that this event will cause damage at installation v_j and the level of damage to be expected at installation v_j. Based on these factors, unidirectional danger factors expressing the amount of danger from one individual installation to another can be calculated. They are referred to as descriptors and can be used to develop a procedure to evaluate domino-effect paths. Such an approach leads to developing a systematic domino-path evaluation procedure in turn leading to optimizing prevention information on the level of every single installation. Identifying the major domino-effect paths can be accomplished by using the approach of enumerating all possible paths in a graph with N nodes and A arcs. This approach offers exact results, but is limited to relatively small problem instances because of exponential increasing computation time. The latter does not pose a problem, since the sub-areas (or domino islands) are supposed to be composed of a relative limited number of installations.

This chapter thus provides recommendations to develop a software tool considering a multi-plant context that is capable of tackling the domino-safety problem as well as the domino-security question. Both problems need to be looked at from an integrating technical perspective. The *modus operandi* suggested is to divide the entire industrial area into smaller sub-areas that are easier to manage with respect to safety and security problems and to investigate precautions in these sub-clusters *vis-à-vis* the continuing of accidental as well as incidental domino effects.

It should be investigated whether dividing the entire multi-plant area into smaller sub-areas can be achieved by eliminating a certain percentage of the installations belonging to the multi-plant area, that is making these installations extremely safe or secure to the extent that they can be regarded as non-existent within the network representing the industrial area. The latter should be based on objective criteria and should be mathematically demonstrable. Since the different sub-areas emerging from this operation based on quantitative data cannot "affect" each other with domino effects, they form isolated domino-effect islands, completing the first objective of the domino-decision-support software.

Furthermore, all installations contributing most to continuing domino effects in the sub-areas need to be "eliminated". To tackle this problem, (i) design basis threat installations should be identified within every sub-area, and (ii) those installations typified by high accidental domino-event continuation characteristics have to be identified within the sub-areas.

6.2.2.2 External Domino-Risk Analysis

External domino-risk analysis can be seen as the process of gathering data and synthesizing information to develop an understanding of the domino risks of a cluster composed of different enterprises. The effort needed to develop this understanding varies depending upon the basic information available for understanding the significance of potential external domino accidents, as illustrated in Figure 6.5.

Only a thorough understanding of the mechanisms and scenarios leading to cascade accidents make it possible for safety and security management to prevent them. However, they are difficult to predict and to manage effectively. Preventative and protective actions should be based on rational qualitative and quantitative information leading to efficient precaution decisions, both for security risks and for safety risks. As already mentioned, a computer-aided tool should be developed by the MPC data administration, allowing ranking installation chains in a complex installations network based on simple installation input data. The results can be

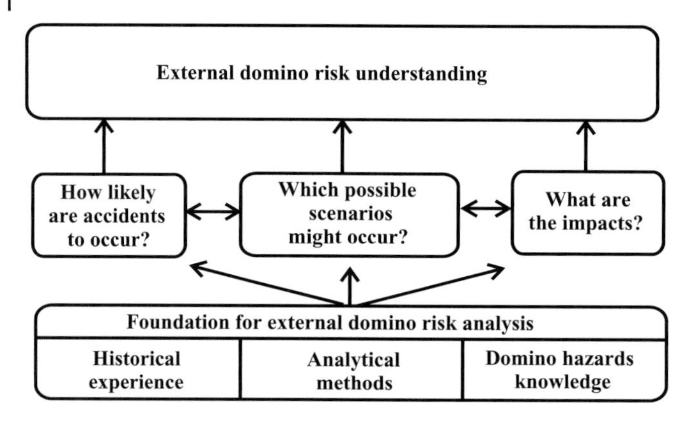

Figure 6.5 Building blocks of external domino-risk analysis (Source: based on CCPS, 2000a).

used by the multi-plant council in the planning of domino precaution measures in the multi-plant area, both for security risks and for safety risks.

6.2.2.3 A Suggested Multi-Plant Domino-Effect-Prioritization Methodology

For prioritizing domino effects in an industrial area, each of the different steps of the twelve-step methodology of Figure 6.6 have to be worked out or on a single-plant level or on a multi-plant level. The consecutive steps are indicated horizontally, while the level at which every stage is performed is shown on the vertical axis of Figure 6.6.

Information on the different steps of the methodology is provided to the participating individual facilities of the multi-plant area. All necessary information about the installation items of the area to be investigated is collected. Further, a "real distances" matrix containing all actual distances between each pair of installations in the area is drafted. These crow-fly distances, for example, calculated from the Lambert coordinates[2] of the installations, can be used. This step is worked out at multi-plant level. At the plant-level effect distances are calculated per installation and per scenario. Installation-failure frequencies are calculated as well. The obtained data are integrated to an aggregated multi-plant level and a matrix of effect distances is drafted for the industrial area of participating companies, containing all possible scenario effect distances between every pair of installations. The real distance between every pair of installations is compared with all possible scenario effect distances from the causing installation to the receiving installation. Based on the difference in distances, a categorization of scenario severity is drawn up into four classes by assigning a scenario domino danger unit (DDU). The data

2) The Lambert projection is one of the most commonly used projections. It produces conformal maps with very low distortion of area, and hence produces maps with some of the lowest overall distortion parameters possible. In fact, it is a standard used by the U.S. Geological Survey, and it is one of the fundamental projections used in the State Plane Coordinate System. Lambert coordinates are often used in safety reports to indicate chemical installation positions.

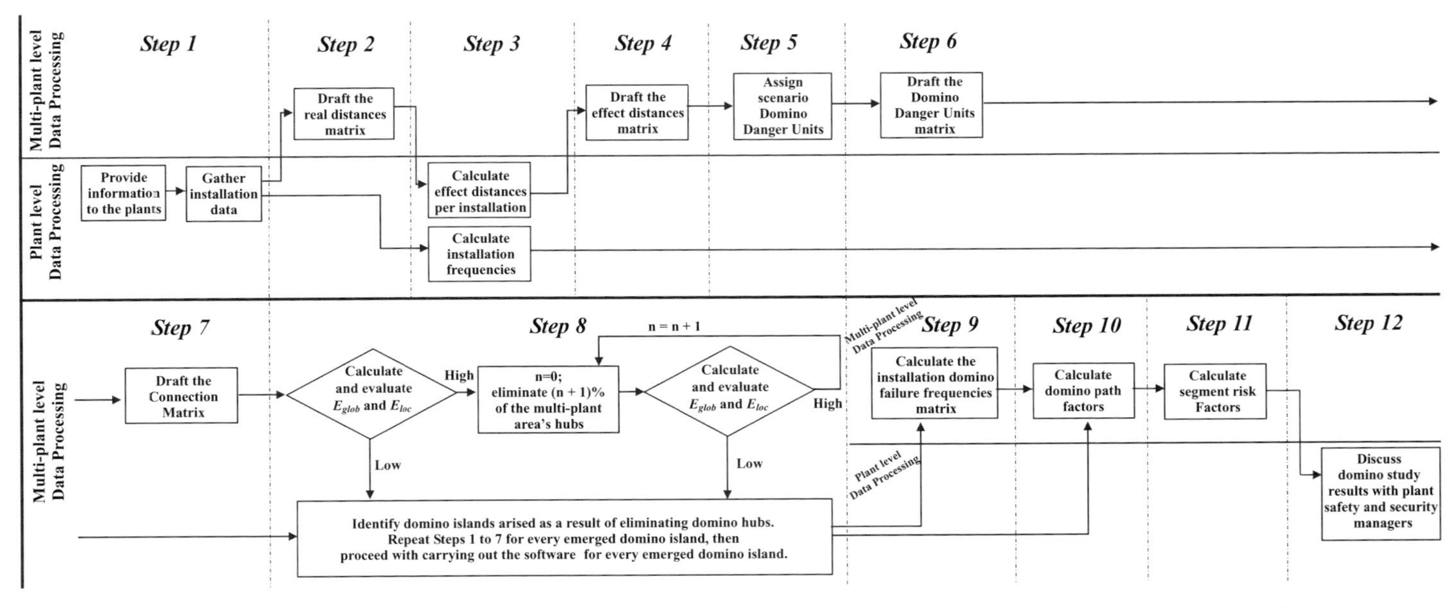

Figure 6.6 Suggested software methodology for ranking installation sequences in a multi-plant context of chemical companies, including safety as well as security risks (to be used by the MPC data administration).

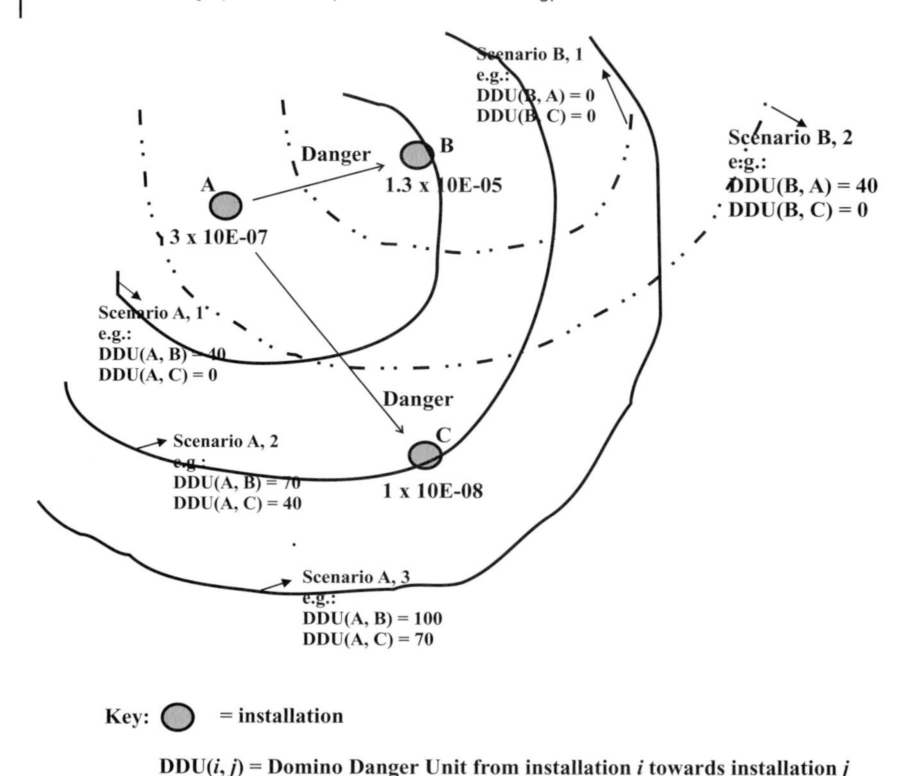

Figure 6.7 Data available at the end of stage nine of the multi-plant domino-effect-prioritization methodology.

available by stage nine for a sub-area (i.e. a domino island) is illustrated in Figure 6.7. The contours displayed in Figure 6.7 are iso-domino-effect curves.

To render the methodology a dynamic tool that takes into account per installation both the outcomes of risk analyses and the preventive and protective measures taken as a result of such analysis procedures, the domino danger unit values or "DDV-values" (see section 6.2.2.4) (e.g., 0, 40, 70 and 100) are entered by the software user. In this way, it is also possible to investigate an industrial area before and after implementing measures to alleviate (internal and) external domino effects. Afterwards, all scenario domino danger units for every unidirectional pair of installations are added to calculate a domino danger unit matrix (step 6).

Stage 7 determines the connection matrix, mapping all existing danger links from every installation to every other installation in the area. The matrix gives a thorough idea of the dangerousness of the area as regards domino effects.

The following stage uses this table of danger links to calculate local and global efficiencies (see section 6.2.2.6) and to verify whether the chemical multi-plant area network is actually a small world. If this is the case, domino islands are identified through simulation of installation hub eliminations within the multi-plant area. Once the isolated sub-networks have been determined, steps 1 through 7 need to be repeated for every sub-area and the remainder of the software (i.e. steps 9 through 12) needs to be carried out (also for every domino island).

Step 9 determines a domino-frequency matrix, listing failure frequencies of couples of installations. These frequencies are determined by multiplying the causing installation-specific failure frequency with the probability of direct ignition of the vulnerable installation, and taking into consideration all vulnerable installations possibly involved in a domino-accident scenario. In step 10 the two matrices are used to calculate domino-path factors, expressing the relative position of the path in a ranking list with respect to the occurrence of domino effects from path begin to path end. Step 11 prioritizes 3-sequence installations of a whole complex multi-plant area by calculating a segment risk factor. In step 12 the results are communicated to, and discussed with, the individual plant-safety managers.

6.2.2.4 Mathematical Approach and Working Procedure for the Suggested Multi-Plant Domino-Effect-Prioritization Methodology

Let $G = (N, A)$ denote a graph of a given network. The network consists of a set of nodes $N = \{v_1, ..., v_n\}$ (e.g., representing chemical installations) and a set of arcs $A = \{a_1, ..., a_m\} \subset N \times N$. Each arc a_k denotes a pair of nodes (v_i, v_j), with $v_i \neq v_j$. It is assumed that $1 \leq N < \infty$ and $0 \leq A < \infty$. The arc (v_i, v_j) is said to be an ordered pair of nodes if it is to be distinguished from the pair (v_j, v_i). If $\forall(v_i, v_j) \in A$ is ordered, (N, A) is called a directed network. The network considered in the suggested domino method is assumed to be directed.

Let s and t be two nodes of $G(N, A)$. A path p from s to t in (N, A) is an alternating sequence of nodes and arcs of the form $p = \langle v'_1, a'_1, v'_2, \ldots, v'_l, a'_l, v'_{l+1} \rangle$, such that:

$$v'_i \in N, \forall i \in \{1, \ldots, l+1\};$$

$$a'_k = (v'_k, v'_{k+1}) \in A, \forall k \in \{1, \ldots, l\};$$

$$v'_1 = s \text{ and } v'_{l+1} = t$$

Nodes s and t are, respectively, called the start node and terminal node of path p. The arc (v'_i, v'_{i+1}) is outgoing from node v'_i and incoming to node v'_{i+1}. Only one arc is allowed between a pair of nodes in the same direction.

A cycle or loop in (N, A) is a path p from one node to itself where all other nodes except s and t are different (that is, $s = t$). A path is said to be loopless *if and only if* all its nodes are different. A null path is a sequence with a single node. In the methodology, paths are assumed to be loopless (since every installation can only be destroyed once).

p_{ij} denotes the path in G from node v_i to node v_j. In many applications involving graphs, it is useful to introduce a variable that measures the weight of each arc, like for example, the arc cost or the arc distance. In the method proposed in this book, the weight of an arc (v_i, v_j) with $v_i \neq v_j$ represents the amount of danger outgoing from installation i onto installation j. Mathematically, the arc weight is simply a scalar (real number) referred to as the domino danger unit (DDU). Let DDU_{ij} denote the weight of an arc (v_i, v_j) with $v_i \neq v_j$, such that:

$$DDU_{ij} \in R^{+} \text{ if } i \neq j$$

$$DDU_{ij} = 0 \text{ if } i = j$$

By calculating all (unidirectional) domino danger units between all nodes in the entire network an installations square danger matrix DDU of order $n \times n$ can be obtained:

$$\mathrm{DDU} = \begin{bmatrix} 0 & \mathrm{DDU}_{12} & \cdots & & \mathrm{DDU}_{1n} \\ \mathrm{DDU}_{21} & 0 & & & \\ \cdots & & 0 & & \cdots \\ & & & 0 & \\ \mathrm{DDU}_{n1} & & \cdots & & 0 \end{bmatrix}$$

In practice, first installation data and real distance data are required for every plant including its various installations (e.g., pressurized storage tanks, cryogenic storage tanks, process vessels, *etc.*). Based on the input data of the installations, the potential effect distances per scenario on the one hand and the failure frequencies on the other hand between every couple of installations should be calculated.

Documents that may be used to perform the calculations are, for example, the Instrument Domino Effects (RIVM, 2003), the Guidelines for Quantitative Risk Assessment (CPR18E, 1999), the Manual of Failure Probability Figures (AMINAL, 2009), *etc.* Of course, documents, instruments, guidelines, information, *etc.* might be country specific and/or multi-area specific.

Figure 6.8 illustrates the working procedure that is used for translating the methodology into programmable code through the use of mathematics. The execution of Step 1A in Figure 6.8 is explained in the next section. The security part of the methodology is explained in the subsequent section. Afterwards, the next – safety-related – stages are clarified. Steps 2 through step 5 indicate the domino danger unit matrix and the failure frequency matrix to be used to prioritize 3-sequence installations in the network.

As already mentioned, the different stages illustrated in Figure 6.8 are explained in the next sub-sections.

6.2.2.5 Developing a Mathematical Model for Carrying Out the *Preliminary Safety and Security Part* of the Suggested Multi-Plant Domino-Effect-Prioritization Methodology

***Step 1A of Figure 6.8*: Domino Danger Units Matrix (DDU_{ij})** The factor DDU_{ij} is a measure of the danger that installation i represents for installation j in terms of domino effects. The domino danger unit DDU_{ij} is calculated between every node v_i and the remaining nodes $v_j(v_j \in N; j \neq i)$ in the graph $G(N,A)$ representing the installations network.

The domino danger unit depends on the distance between the two installations i and j. The effect-distance connected with a possible accident scenario from one

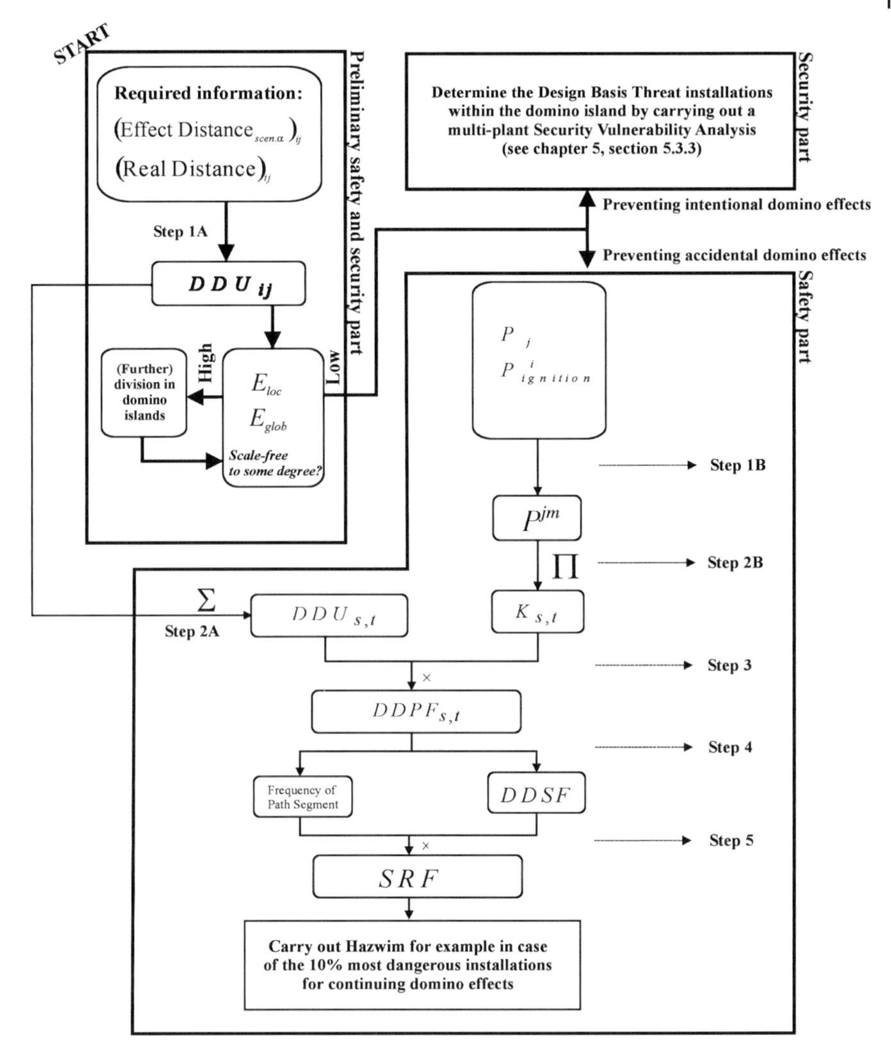

Figure 6.8 Suggested software working procedure for investigating domino risks in multi-plant areas.

installation to another is linked to the real distance between the two installations involved. Depending on the difference in distance, a domino distance factor (DDF) can be defined, as expounded later in this section. Such a strategy allows for a relative ranking of the installation pairs in the network to determine whether they are more or less dangerous (as regards domino effects) than the other installation pairs. The philosophy behind this approach is to establish the relative importance of couples of installations from a domino safety and security viewpoint, before performing additional and much more costly domino-hazard-evaluation studies.

In general, there are several possible scenarios responsible for domino accidents. Khan and Abbasi (2001) suggest that possible scenarios for initiating domino effects include pool fire, flash fire, fireball, jet fire, vapor-cloud explosion (VCE), unconfined vapor-cloud explosion, boiling liquid expanding vapor explosion (BLEVE), *etc.* An excerpt of a domino distance table from the instrument domino effects (RIVM, 2003) offering domino scenario effect distances for the case of two quantities (i.e. smaller than 5 ton or between 5 and 10 ton of a flammable gas) stored in bulk or processed, in both cases under pressurized conditions, is shown in Table 6.1.

If, for example, 8 ton butane (which is a substance characterized by the label *GF2*) is stored under pressurized conditions, 5 scenarios caused by instantaneous failure of the storage tank or the process installation (e.g. called "installation v_i") could give rise to a secondary domino event: an overpressure wave (Sc.1), missile projection (Sc.2), a boiling liquid expanding vapor explosion (BLEVE) (Sc.3), a vapor-cloud explosion (VCE) (Sc.4), and a pool fire (Sc.5). Depending on the protection of the secondary installation labeled v_j (e.g., protected against overpressure waves up to 0.3 bar and not protected against heat radiation), domino distances are given in Table 6.1 between both installations (from v_i to v_j): 23 m (Sc.1), 296 m (Sc.2), 65 m (Sc.3), 205 m (Sc.4), and 130 m (Sc.5).

The effect-distance[3] associated with a possible accident scenario[4] from one installation to another is linked to the real distance between the two installations concerned. Depending on the difference in both distances (real distance and effect distance), a domino distance factor (DDF) can be defined, using four possible categories. These numerical values represent the relative level of importance given to the pairs of installations with respect to their danger for inducing or continuing domino effects. Such a strategy allows for a relative ranking of the installation pairs in the network to identify the most dangerous ones.

The choice of the numerical DDF values (giving rise to the four different categories) is actually a preference decision problem for safety and security management wishing to apply the tool. The set of possible (mutually exclusive) alternatives from which safety and security management must choose, balances the consequences of accident scenarios. For example, the outcome of one major accident scenario is compared with the consequences of two accident scenarios ranked in one category below. Hence, a preference relationship has to be fixed by the tool user as regards the consequences of domino-event scenarios. As an example, the method's scenario preference relationship, denoted by $\succ$,[5] may be set to be:

the consequences of two accident scenarios $\succ$ *the consequences of one accident scenario ranked one category higher*

3) The distance at which an accident (scenario) might have an impact.
4) In general, there are several scenarios possible for causing domino accidents. Possible scenarios for initiating domino effects include overpressure effects, missile projection, pool fire, flash fire, fireball, jet fire, VCE, and BLEVE.
5) We read $x \succ y$ as "x is preferred to y".

Table 6.1 Pressurized bulk storage/pressurized processing: domino effect distances (in m.) for different accident scenarios.

Msys (ton)	Bursting/peak overpressure (bar)				Missiles	BLEVE ($R_{fireball}$)	VCE/peak overpressure (bar)					Pool fire/heat radiation (kW/m^2)		
	0.1	0.2	0.3	0.45			LFL-Distance	0.1	0.2	0.3	0.45	D_{max} (m)	8	37.5
≤5														
GF1	30	21	16	11		52	40	141	90	72	58	24.8	67	39
GF2	35	23	18	16			120	204	168	157	150	23.9	97	54
GF3	61	37	27	21			115	187	146	138	135	–	–	–
>5–10					296									
GF1	37	26	20	14		65	55	177	113	90	73	35.0	89	53
GF2	43	29	23	16			157	266	220	205	195	33.4	130	74
GF3	77	47	34	27			134	237	172	161	155	–	–	–

Source: Instrument Domino Effects (RIVM, 2003).

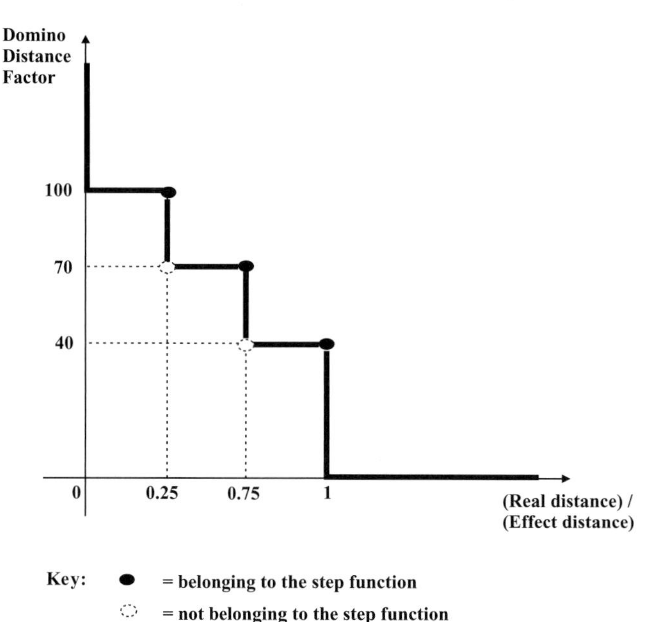

Figure 6.9 Domino distance factor for ranking the domino danger between two installations.

The exemplary preference translates into the following choice of DDF values. For the exemplary parameter set of the tool, for a specified scenario, if the real distance between both installation items does not exceed a quarter of the theoretical effect-distance, the domino distance factor equals 100. On the other hand, if the real distance strictly exceeds the effect-distance, DDF = 0. In the case where the real distance strictly exceeds one quarter of the effect-distance and is lower than three quarters of the theoretical effect-distance, DDF = 70. In the final case where the real distance is restricted by the effect-distance and strictly exceeds three quarters of the effect-distance, DDF = 40. Figure 6.9 illustrates this exemplary parameter setting, which should be easily modifiable by the user.

Consider the example of pressurized butane storage (installation v_i) of Table 6.1. In this case, Figure 6.10 illustrates the calculated domino distance factors between installation v_i and installation v_j.

A domino danger unit is defined to express the escalation dangerousness from one installation to another, by summing the DDF values for the different possible scenarios (1, ...,*K*) outgoing from installation v_i and incoming to installation v_j:

$$\mathrm{DDU}_{ij} = \sum_{\text{scenario } k=1}^{\text{scenario } K} (\mathrm{DDF}_{ij})_k \qquad (6.1)$$

Hence, for the example of butane storage, $\mathrm{DDU}_{ij} = 0 + 0 + 40 + 70 + 70 = 180$.

This formula (5.1) is used to calculate the weights of the directed arcs (v_i, v_j) and (v_j, v_i) between every pair of installations. A total danger TD_i for every node v_i can be calculated by summing the DDU_{ij}:

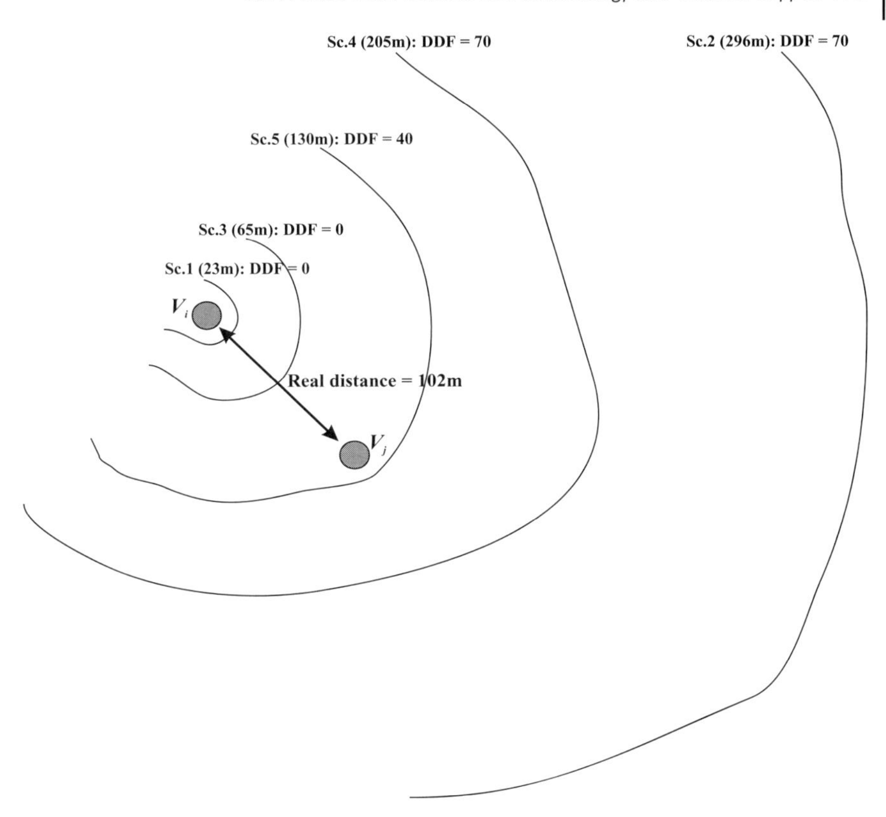

Figure 6.10 DDF calculation example.

$$\mathrm{TD}_i = \sum_{j(\neq i)\in G} \mathrm{DDU}_{ij} \tag{6.2}$$

In the butane case, if the industrial area consists of three installations (v_i, v_j, and v_k) and if, for example, the domino danger unit between v_i and v_k is 110, then $\mathrm{TD}_i = 180 + 110 = 290$ (see Figure 6.11).

Installations characterized with large TD values can in fact be regarded as domino hubs within the industrial network. Such installations thus represent the highest danger for initiating and/or continuing devastating domino effects within the area. Hence, by extracting the nodes displaying the highest danger connectivity in a chemical installations' network (i.e. by eliminating the domino hubs characterized with high TD values) the domino danger interconnectedness within the network is rearranged and its global and local danger is substantially lowcrcd.

However, this is the standard static situation in which the level of prevention and protection as a result of (safety and security) risk analyses carried out on the causing installation has not been considered. In order to make the toolkit more dynamic and hence more realistic, a risk-analysis factor can be introduced.

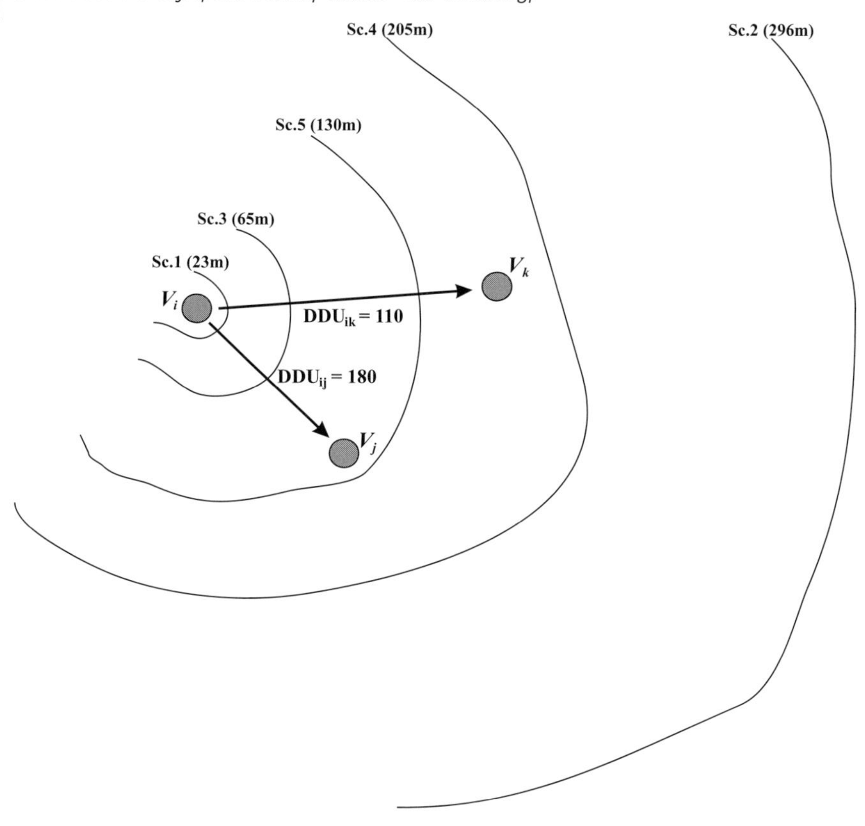

Figure 6.11 Calculation of TD_i for an industrial area composed of three installations.

The suggested methodology/tool should allow the user to insert his/her own four levels of DDFs correlating with a particular risk-analysis factor category, that is 1, 2, 3 or 4. Using the risk-analysis factor ranging from 1 to 4, the standard domino distance factors are automatically updated by the software depending on the software user's input. As already mentioned, adjusting DDF values of installations over time might for example, be based on previous risk analysis performances. The exemplary domino distance factors of 100, 70, 40 and 0 for example correspond to a risk-analysis factor of category 1. Other possible values corresponding with a different risk-analysis factor can be used. To determine the most appropriate risk-analysis factor values, a sensitivity analysis should be performed using a variety of domino distance factor values. The obtained installation prioritization results should not be affected by changing the DDF values, indicating that the results using these DDF values may be seen as robust.

6.2.2.6 Developing a Mathematical Model for Carrying Out the *Security Part* of the Suggested Multi-Plant Domino-Effect-Prioritization Methodology

The emerging importance of so-called *scale-free networks* is a direct consequence of the widespread and multidisciplinary real-life occurrence of such networks, for example, the world-wide web, neural networks, social networks, sexual relationships, airline routes, connections between actors in Hollywood, terrorist networks, protein-interaction networks, metabolic networks of life-sustaining chemical reactions inside cells, *etc.* (Barabasi, 2003; Buchanan, 2003; Gong and van Leeuwen, 2003). The importance of such networks and its application to an external domino-effect-prevention tool is discussed in this section.

To diagnose a network as scale-free, it has to be typified by three main characteristics (Alberich *et al.*, 2005), that is (i) every pair of nodes can be connected through a short path within the network, (ii) the probability that two nodes are linked is greater if they share a neighbor, and (iii) the fraction of nodes with k neighbors decays roughly as a function of the form $k^{-\gamma}$ for some positive exponent γ.

A network satisfying only features (i) and (ii) is called a *small-world network* (Watts, 2004). The third characteristic, so typical of *scale-free* networks, can be defined as the network obeying a power-law distribution in the number of connections between nodes in the network, that is very few nodes exhibit very high connectivity (such nodes are generally called "hubs") while the vast majority of nodes have only few links.

Barabasi (2003) and coresearchers have found that the mathematical properties of scale-free networks have a significant impact upon their stability. Scale-free networks are extremely tolerant of random failures; information exchange scale-free networks for example, can absorb random failures in up to 80% of their nodes before they breakdown. However, these researchers also indicate attacks simultaneously destroying as few as 10% to 15% of such scale-free network's hubs can cause the entire network to disintegrate.

As ever more networks are identified to be scale-free the question arises whether an industrial area composed of a large number of chemical installations might be considered as a scale-free network. If so, the question of the network possibly collapsing can be rephrased and it may be investigated whether chemical areas might be protected in a more rational way. Could it be possible that, by taking adequate precautions with possibly very few installations, an industrial area would be much better secured against the propagation of domino effects? If the latter is the case, a more mathematical approach to carrying out domino-risk analyses in an industrial area might be appropriate from a security viewpoint.

It must be borne in mind that if safety management fails to systematically carry out analyses and assessments to identify and to evaluate an installation's risks, its danger towards the other installations in the area automatically increases (over time). The latter can be explained considering the aging of equipment and of materials as well as human nature being likely to lower safety attention over time when no incident occurs. This phenomenon is known as entropic risk (Mol, 2003). The automatic increasing of danger in case of no risk assessments being carried out can also be referred to as the "dynamic behavior of a chemical installation". To study the dynamic properties of an industrial area, the area needs to be modeled as a network of installations. The consequences of such dynamic features for the network depend largely on the network topology and its structure. In order to address these issues, three research questions to gain

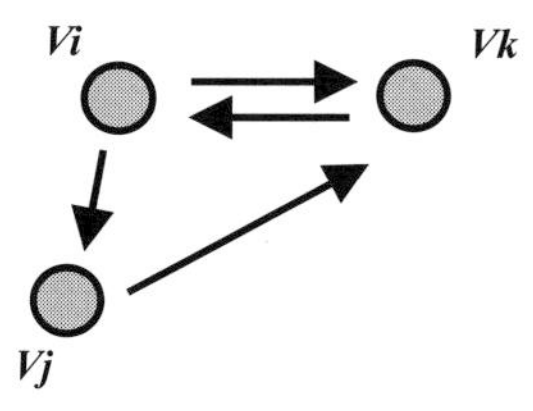

Figure 6.12 An illustrative network of three nodes.

a better understanding of the complex environment of a chemical cluster can be investigated:

i) Can an industrial area be modeled as a structured network, that is a directed weighted graph?
ii) Can an industrial area be considered as a *small-world*?
iii) Does an industrial area display *scale-free* characteristics?

The first research question (i) was treated in the previous section and is thus positively answered. How to mathematically investigate the second and the third research questions (ii) and (iii), respectively, is explained hereafter.

The mathematical translation of the three characteristics defining a scale-free network, is based on the evaluation of three quantities, that is the characteristic path length $L = \frac{1}{N(N-1)} \sum_{i \neq j} d_{ij}$, the clustering coefficient $C = \frac{1}{N} \sum_{i} C_i$, and the power-law degree[6] distribution $P(k) \sim k^{-\gamma}$. It should be noted that $\boldsymbol{d}$ is the matrix of the shortest path weights (the smallest sum of the weights on the edges throughout all the possible paths in the graph, whatever the connotation of the weights) between two generic vertices v_i and v_j. C_i is the number of edges existing in G_i, the sub-graph of the connected neighbors of v_i, divided by the maximum possible number $k_i \cdot (k_i - 1)$, with k_i the number of edges originating from vertex v_i.

For a better comprehension of the latter terminology, consider the example illustrated in Figure 6.12. In the example of Figure 6.12, $k_i = 2$, $C_i = 1/2$.

The parameters L and C can be used to investigate whether domino effects might proceed efficiently in the network, globally and/or locally. To this end, the "efficiency" parameter E of a network introduced by Latora and Marchiori (2001, 2002, 2003) can be utilized. Efficiency allows for small-world networks to be seen as systems that are both globally and locally efficient. $1/L$ and C can be seen as first approximations of E, evaluated on a global and a local scale respectively. Small-world networks are highly clustered and possess small characteristic path lengths. Small-world networks, which are scale-free, have a distribution of the number of connections of a generic vertex v_i following a power law.

By using L and C, or E on a local or global scale, a network's security behavior relative to another network's security behavior can be studied, and, for example, domino effects design basis threats may be decreased by improving security coun-

6) The degree of a vertex is the total number of its outgoing links.

termeasures by carrying out simulations or maximum global and local efficiencies within industrial areas can be recommended (recommendations resulting from comparative studies of industrial clusters).

As already mentioned, the domino threat that every vertex (installation) poses to every other vertex (installation) in the network can be represented by the domino danger unit associated with the directed edge between every couple of vertices. On the one hand, the efficiency ε_{ij} of the network is then assumed to be proportional to the aggregated danger of the path between two nodes v_i and v_j. On the other hand, the number of links is inversely proportional to the efficiency ε_{ij}. The more links there are to get from one installation to another in the network, the less dangerous the network is, thus the lower the efficiency according to the measures of Latora and Marchiori.

The "direct efficiency" can then be calculated by considering the path in a network between two nodes v_i and v_j that is characterized by the lowest number of steps to go from v_i to v_j. Hence, the direct efficiency can be expressed in mathematical terms as follows:

$$\varepsilon_{ij}^{\text{direct}} = \underset{\text{path}i \to j}{\text{Min}} \left(\frac{\sum_{i' \neq j' \in G} \text{DDU}_{i'j'}}{A_s^{ij}} \right), \forall v_i, v_j \tag{6.3}$$

with A_s^{ij} = number of links of the path going from v_i to v_j, and having the lowest number of links. It is assumed that $A_s^{ij} \geq 1$. If no path exists between v_i and v_j, $A_s^{ij} = +\infty$ and thus consistently $\varepsilon_{ij} = 0$.

The "indirect efficiency" of the network can be determined by calculating the maximum aggregated danger of the path between two nodes v_i and v_j. To this end, a path with highest aggregated weight between the nodes v_i and v_j (and using the domino danger units as weights) for every couple of nodes in the graph, has to be constructed. Since the number of links of the longest path (from node v_i to node v_j having the highest aggregated danger), A_l^{ij}, is inversely proportional to the efficiency ε_{ij}, this factor also has to be taken into account. Hence, the indirect efficiency can be expressed in mathematical terms as follows:

$$\varepsilon_{ij}^{\text{indirect}} = \underset{\text{path}i \to j}{\text{Max}} \left(\frac{\sum_{i' \neq j' \in G} \text{DDU}_{i'j'}}{A_l^{ij}} \right), \forall v_i, v_j \tag{6.4}$$

with A_l^{ij} = number of links of the path going from v_i to v_j, and having the highest aggregated danger. It is assumed that $A_l^{ij} \geq 1$. If no path exists between v_i and v_j, $A_l^{ij} = +\infty$ and thus consistently $\varepsilon_{ij} = 0$.

The average efficiency of graph G can be defined as (Smith, 1988):

$$E(G) = \frac{\sum_{i \neq j \in G} \varepsilon_{ij}}{N(N-1)} \tag{6.5}$$

The global efficiency E_{glob} of G is the efficiency of the entire network and can be defined as:

$$E_{\text{glob}} = \frac{E(G^{\text{direct}})}{E(G^{\text{indirect}})} \tag{6.6}$$

A high global efficiency is consistent with a low characteristic path length (Latora and Marchiori, 2003).

Since the efficiency can also be evaluated for any sub-graph of G, it can be used to evaluate the local properties of the graph. Therefore, using the efficiencies for every graph G_i, the local efficiency of G can be expressed:

$$E_{\text{loc}}(G) = \frac{1}{N}\sum_{i \in G} \frac{E(G_i^{\text{direct}})}{E(G_i^{\text{indirect}})} \tag{6.7}$$

with:

$$E(G_i^{\text{direct}}) = \frac{1}{k_i(k_i-1)} \sum_{l \neq m \in G_i} \underset{\text{path}l \to m}{\text{Min}} \left(\frac{\sum_{i' \neq j' \in G_i} \text{DDU}_{i'j'}}{A_s^{lm}} \right) \tag{6.8}$$

and:

$$E(G_i^{\text{indirect}}) = \frac{1}{k_i(k_i-1)} \sum_{l \neq m \in G_i} \underset{\text{path}l \to m}{\text{Max}} \left(\frac{\sum_{i' \neq j' \in G_i} \text{DDU}_{i'j'}}{A_l^{lm}} \right) \tag{6.9}$$

Since $v_i \notin G_i$, E_{loc} indicates the fault tolerance of the system, that is the efficiency of the domino danger spreading between the first neighbors of v_i when v_i is removed.

In general, a network $G(N, A)$ representing an industrial chemical area can then be identified as being a small-world if it gives rise to both a high E_{glob} and a high E_{loc}, such a network thus proving efficient both in global and local domino danger spreading.

It should, however, be noted that these values for E_{glob} and E_{loc} are not absolute values, but they should be interpreted as relative cluster-related values providing an idea of the "domino security dangerousness" of chemical clusters. Hence, it is not possible to determine an industrial area unambiguously to be "a small-world". Using these efficiency values it is possible to rank industrial areas with respect to their small-world behavior and thus make a prioritization as regards their security countermeasures' needs. Further research for refining and validating the method, is being carried out.

Summarizing the first point of interest, that is the small-world features of the cluster network, the industrial area is considered as a graph $G(N, A)$ and is represented by its connection matrix[7] $\boldsymbol{a}$ and by its domino danger units matrix **DDU**. Subsequently, code can be used to determine the global efficiency and the local efficiency of the network. If both calculated efficiencies return high values, the network (i.e. the chemical installations cluster) displays small-world behavior. First, the network is only $(100 - E_{\text{glob}})$% less efficient in propagating domino effects than the ideal network with a danger link from each installation to the others.

7) $\boldsymbol{a}$ is the $N \times N$ matrix where $a_{ij} = 1$ if there exists an edge (v_i, v_j) and $a_{ij} = 0$ otherwise.

Second, a high local efficiency of E_{loc} indicates that the site is very "fault tolerant": removing an installation chosen at random in the area, and all the danger links with it, will not dramatically affect the efficiency in the danger linking between the neighbors of that installation.

A second point of interest is the investigation of the power-law distribution of the danger connection topology of the network(s). The installations situated on an industrial area (or sub-area) might be characterized by an uneven distribution of danger connectedness. Some (very dangerous) installations then act as "highly connected" hubs for initiating or continuing domino effects. This might dramatically influence the way the industrial area network breaks down. Hence, to determine whether an industrial area is scale-free to some extent, the MPC should investigate to what extent the dangers linking the installations to a network follow a power-law distribution. To investigate the latter in a directed weighted graph, the total domino danger TD_i of a vertex is used to obtain a vertex "out-strength distribution". In the case of an industrial area, the automatic increase of the TD_i (if no risk analyses are carried out and no resulting precaution and protection measures are taken), has a higher impact on already dangerous installations since their domino danger increases towards a lot of nearby installations. Therefore, a network $G(N, A)$ representing an industrial chemical area can be identified as being scale-free to some degree if it is typified by an out-strength distribution following a power law to some degree.

Once the investigated industrial area is classified by the multi-plant council data administration to be a network displaying scale-free characteristics, the network (i.e. the industrial cluster) should be divided into smaller networks, which in turn are to be investigated for their uneven distribution of danger connectedness (see Figure 6.6). To this end, the impact of eliminating domino hubs in the installations' network is studied by removing the hubs one per cent at a time, starting with the hubs having the highest TD value. For every step, the network is recalculated (i.e. the number of edges as well as the total danger of the new network are determined and the structure of the new network is drafted). Using this simulation technique, the emerging domino islands (i.e. smaller areas in the network forming networks of their own with no links in between) can be observed and investigated. Table 6.2, which is the result of a case study, illustrates the latter process, showing that by eliminating as less as 1% of the domino danger hubs in the area, about 20% of the total domino danger and about 12% of the number of links are simultaneously eliminated. Moreover, the whole area is divided into three domino islands where no domino effects can occur in between. Table 6.2 further demonstrates that by eliminating as few as 3% of the domino danger hubs in the area, about 30% of the total domino danger and about 20% of the number of links were simultaneously eliminated in this particular real-case industrial area of two adjacent chemical plants. Moreover, the whole industrial area was divided into four domino islands where no domino effects can occur in between. As a result, the global danger for inducing domino effects in this particular industrial area decreased substantially. Furthermore, Table 6.2 clearly demonstrates that if ever more domino danger hubs are being removed, ever

Table 6.2 Overview of characteristics of the different networks emerged by eliminating domino hubs from the original real-case industrial area network.

Reduction of highly connected installations	Number of installations	Number of links between installations	Additional reduction of links	Total amount of network domino danger	Additional reduction of domino danger	Number of domino islands
0	227	5244	–	488 410	–	1
−1%	224	4625	−11.80%	394 170	−19.29%	3
−3%	220	4157	−8.92%	340 400	−11.01%	4
−5%	215	3675	−9.19%	292 330	−9.84%	4
−7%	211	3365	−5.91%	262 930	−6.02%	5
−9%	206	2978	−7.38%	236 310	−5.45%	5
−11%	202	2715	−5.01%	216 510	−4.05%	5
−13%	197	2358	−6.81%	189 670	−5.50%	6
−15%	192	2069	−5.51%	169 660	−4.10%	7

more disjoined domino islands emerge in the industrial area. Further removing domino danger hubs in the area thus reveals additional domino islands and further decreases the number of links and the total domino danger, compared with the original network. The higher the percentage of hubs removed, the lower the additional reduction of domino links and of the total domino danger. As a result, the global danger for inducing domino effects in the industrial area ever more decreases.

The multi-plant council data administration thus has to make a trade-off between security efforts (costs) and the preferred number of emerged domino islands.

Within every domino island, asset-based and scenario-based SVAs should further be carried out to pinpoint the design basis threat installations of the multi-plant area.

A weakness of the presented method seems to be that frequencies or probabilities of the initiating events are not included in the application. However, dealing with security risks should indeed be independent of frequencies and probabilities, taking only the potential consequences of designed domino effects into account, as terrorists are not concerned with coincidence at all. The proposed security part of the methodology is thus purely deterministic and merely takes the outcome of intentional escalation effects into account.

Furthermore, it should be noted that the proposed security software application identifies those installations that should be better secured against potential threats in order to substantially decrease the possibility of a deliberate domino event in the area, but it does not consider the question of *how to* eliminate those domino danger hubs. It is up to safety and security management of the industrial complex to discuss the required countermeasures (see also Section 5.3.3).

6.2.2.7 Developing a Mathematical Model for Carrying Out the *Safety Part* of the Suggested Multi-Plant Domino-Effect-Prioritization Methodology

The sub-area network problem for identifying those installations contributing most to domino continuation within the sub-area can be tackled by repeatedly enumerating all possible paths in every smaller sub-sub-area combination (consisting of, for example, five nodes) resulting in a list of all possible paths consisting of two to four domino events. Such an approach is justified, because literature research on major accidents where domino effects are involved has revealed that it is very difficult and often impossible to discern and to analyze domino effects with cardinality greater than three, that is accidents consisting of more than four successive domino events (see e.g., Fievez, 1996; Lees, 1996). Another important argument is that the propagation of domino events should be stopped as early as possible in a chain in order to keep the consequences of the accident under control as effectively as possible. For these reasons, the constraint can be made that not more than four successive domino events are allowed to happen. Given a directed graph $G = (N, A)$ with $N \geq 5$, all possible combinations of 5-node sub-graphs[8] are to be listed. In the first step, all feasible paths in order of decreasing length between each pair of nodes in the sub-graphs are enumerated, and in the second stage, the list obtained in the first step is used to enumerate and to rank all featuring 3-node sequences according to the risk they represent in initiating or propagating catastrophic domino accidents. In the case of the domino-continuation problem in particular, the enumeration solution seems to be the best choice. By reducing the original (possibly large) network to a number of 5-node graphs, analyzing and evaluating all featuring 3 node sequences in very large clusters and/or sub-clusters is made possible.

Sequences (or segments) consisting of three installations where two successive domino effects may occur, represent easy parameters in understanding locations in the network where danger in terms of domino accidents is the greatest. Investigating more installations per sequence (for example by considering 4-node segments) is more difficult for both generation and interpretation of results[9] and thus sub-optimal for decision taking, whereas 2-node segments are already treated by other software and fail to capture higher-order domino accidents.

In the mathematical approach proposed in this book every possible sub-graph combination consisting of five nodes is systematically analyzed. Subsequently, the occurrence probability of every enumerated "domino path" of each small network is determined. Taking the overall domino danger and the overall probability of a path into account, an expected path danger as regards domino effects can be calculated. A classification can be made using the occurrence frequency of 3-node path segments as one factor and the sum of all path dangers encompassing 3-node path segments as another factor. Such a ranking is important because safety managers have to make pro-active priority choices in the most efficient way when

8) A sub-graph is a graph that in its entirety forms part of a larger graph.
9) The 4-node sequence result leads for example to the question which installation to protect first: the second or the third of such a segment.

deciding where to take safety measures. All possible domino paths between any source node s and a chosen terminal node t ($s \neq t$) in such a sub-network, are enumerated. The accumulated domino danger per path is calculated. By then examining every feasible 3-node path sequence out of every enumerated domino path, conclusions can be drawn about domino-chain accidents. A perception of the most dangerous path segments of two adjacent arcs in a complex installations network of a multi-plant area is acquired.

Figure 6.8 shows the different stages for carrying out the safety part of the proposed methodology. The mathematical elaboration of the different steps forming the approach illustrated in Figure 6.8 is given hereunder.

***Step 1B of Figure 6.8*: Installation Probability Factor (P^{jm})** Many methods for the assessment of accident propagation in the literature are based on the identification of threshold values for the primary physical effects. However, different values are suggested by different sources and hence the reliability of these thresholds is questionable. A solid quantitative assessment requires the estimation of the probability of propagation of which are the most promising propagation functions based on probit functions (Cozzani and Salzano, 2004). However, the few probabilistic models available for propagation probability are not always consistent in the output of results. Therefore, reliable research leading to accurate and standardized quantitative calculations is still needed in the field.

Nevertheless, for ranking purposes scenario probability data can be regarded as ordinal data and therefore can be used. For example, data for assessing the probability of failure for the different installation types existing in the chemical processing industry should be used in the methodology.

Consecutive events being part of a domino accident are assumed to be independent occurrences. In this way the probabilities of such events happening due to a successive combination of dangerous installations can be modeled in a simplified way. The latter assumptions are permitted because the goal of the study is to rank installation sequences relatively. Hence, ordinal data are used implying only order, for example, "most" to "least" or *vice versa*. Therefore, these failure frequency data can be used to consistently and objectively calculate domino-path frequencies. It must be stated that a conservative independent occurrence approach can be a clear overestimate of the "real" domino-chain probability. However, since the overestimate is consequently maintained, the ranking results also prove consistent.

Let the factor $P_{i,j}$ represent the probability that a domino event will occur on equipment i *(being considered in isolation)* caused by a *direct domino effect from installation* j (Atkins, 1998). A formula for calculating the probability of a domino event initiated by a particular installation j situated in a surrounding of $(N - 1)$ installation items must bc identified.

To avoid calculating the cross-products explicitly, it is possible to convert the actual probability of a domino accident occurring, to the probability of the success state. The probability of a domino event not arising as a result of primary failure of equipment j on item i can thus be expressed as $1 - P_{i,j}$. If terms $1 - P_{i,j}$ for each of the $(N - 1)$ installations are calculated and are multiplied, then the probability of a domino event not arising as a result of primary failure of equipment j on items 1 to $(N-1)$ becomes

$$\prod_{i=1}^{N-1}(1-P_{i,j}) \tag{6.10}$$

The probability of a domino accident arising from installation j in a network of N chemical installations (including j) is thus

$$1-\prod_{i=1}^{N-1}(1-P_{i,j}) \tag{6.11}$$

This expression gives the overall probability of a domino accident occurring from a particular installation item.

To make a distinction between the different types of installations that might be affected by the causing installation j, the contribution of a particular vulnerable installation i to the likelihood of a domino accident caused by installation j has to be determined. Assume a network consisting of four chemical installations: A, B, C and D, and let A be the initiating installation item. Assume that the installation B is the first to fail by failure of A. The conditional probability of A giving rise to a domino accident is $1-(1-P_{B,A})$. The conditional probability of the second installation to be affected, for example, C, is the combined probability of the first and second installation items minus the probability of the first installation item, that isi.e. $\lfloor 1-(1-p_{B,A})(1-P_{C,A})\rfloor - \lfloor 1-(1-p_{B,A})\rfloor$. A simplified generalized expression for installation i being the mth installation to be hit by the initiating installation j can be derived:

$$P^{jm}=\prod_{i=1}^{m-1}(1-P_{i,j})-\prod_{i=1}^{m}(1-P_{i,j}) \tag{6.12}$$

This expression can be written as

$$P^{jm}=P_{m,j}\prod_{i=1}^{m-1}(1-P_{i,j}) \tag{6.13}$$

Expression (6.13) can be used to make a difference between the installations that can be affected by the causing installation.

If the domino effect distance of any scenario at j is less than the real distance between j and i, the scenario probability $P_{i,j}$ is zero; otherwise $P_{i,j}$ can be calculated by multiplying the failure probability P_j of the installation j with the probability of direct ignition for stationary installations or for transport units in an establishment P^{j}_{ignition}.

If some scenario occurs at installation j, first the effect distance of the scenario is compared with the real distance between installation j and installation i. Secondly, if an escalation at installation i is possible, the probability that an ignition will occur at installation i must be considered as well. Considering the latter assumptions, expression (6.13) can be written as

$$P^{jm} = P_j P^j_{\text{ignition}} P^m_{\text{ignition}} \prod_{i=1}^{m-1} \left(1 - P_j P^j_{\text{ignition}} P^i_{\text{ignition}}\right) \tag{6.14}$$

with:

P^{jm} = resulting probability of the mth installation possibly affected by a failure of installation j;

P_j = generic failure frequency of installation j for instantaneous escalation accident per unit of time;

P^i_{ignition} = probability of direct ignition at installation i.

***Step 2A of Figure 6.8*: Domino Danger Unit of a Path** Given two nodes x and y of a path p_{ij}, $\text{seg}_{xy}^{p_{ij}}$ is defined as a segment of path p_{ij} if it coincides with p_{ij} from x until y. The accumulated $\text{DDU}_{s,t}$ of a path $p = \langle s, a'_1, v'_2, \ldots, v'_l, a'_l, t\rangle$ is obtained by summing the individual DDU_{ij} of the 2-node segments, that is arcs, of the path. Thus,

$$\text{DDU}_{s,t} = \sum_{\forall ij = i(i+1) \in \text{seg}_{xy}^{p_{s,t}}} \text{DDU}_{ij} \tag{6.15}$$

***Step 2B of Figure 6.8*: Probability of a Path ($K_{s,t}$)** Assuming the independence of successive accident events (leading to conservative calculation results), the probability of a feasible domino path occurring in an installations network can be calculated as follows:

$$K_{s,t} = P^{sv'_2} \cdot P^{v'_2 v'_3} \cdot \ldots \cdot P^{(v'_l - 1)v'_l} \cdot P^{(t-1)t} \tag{6.16}$$

***Step 3 of Figure 6.8*: Domino Danger Path Factor ($\text{DDPF}_{s,t}$)** For obtaining the risk level of a possible domino path, a factor representing the consequences of the path accident, that is, the $\text{DDU}_{s,t}$, has to be multiplied with a factor representing the path accident probability, in this case $K_{s,t}$. Therefore, the $\text{DDPF}_{s,t}$ is determined by multiplying the overall domino danger unit of the path and the overall probability of the path:

$$\text{DDPF}_{s,t} = \text{DDU}_{s,t} \cdot K_{s,t} \tag{6.17}$$

For every possible domino path p in the chemical industrial area, a domino danger path factor (DDPF) is calculated.

***Step 4 of Figure 6.8*: Domino Danger Segment Factor (DDSF) and the Path Segment Frequency** The segment frequency is obtained by counting all the paths encompassing the 3-node segment in the entire installations network, whereas the overall DDSF is obtained by summing all DDPFs of the paths encompassing the segment in the network.

***Step 5 of Figure 6.8*: Segment Risk Factor (*SRF*)** The segment risk factor is calculated by multiplying the frequency of the path segment with the overall DDPF of the path segment or the domino danger segment factor (DDSF).

The safety part of the proposed methodology should result in a clear, objective, relative ranking (using the SRF) of priority installations where precautions need to be taken. Relative ranking is an analysis strategy that allows for a comparison of the attributes of several items, *in casu* installation sequences. By using relative ranking in the case of domino effects, information is provided regarding any alternative that appears to be the most dangerous sequence with respect to initiating or continuing domino events. Comparisons are based on numerical values that represent the relative level of danger that is given to each pair of installations. This approach can be applied in the example of an existing situation within a chemical cluster of installations to pinpoint the installations where caution is most needed for the prevention of safety-risk-related domino effects. To compare pairs of industrial installations in terms of their domino danger contribution, quantitative data have to be used. For this purpose, all different installation types have to be linked not only to the outcome of possible domino accident scenarios (as in the deterministic domino security part of the methodology), but also to scenario failure frequency data (as in the probabilistic domino safety part of the methodology).

After determining the most dangerous installations (for continuing domino effects from a safety viewpoint) within the multi-plant area, these installations need to be investigated by a risk-analysis technique. If the installations may affect different companies within the multi-plant area, the area can be examined using the hazwim framework (see Chapter 5, Section 5.2.2.3). Otherwise, single-company safety management should follow up the pinpointed installation(s). Due to their popularity among risk analysts in the chemical industry, hazop analysis, what-if analysis and the risk matrix can be recommended as risk-analysis techniques.

In fact, the hazop procedure and the what-if procedure may be adapted to investigate domino hazards. Hazop users apply so-called “guidewords”. The original guideword-based hazop approach was developed by ICI Chemicals, in the mids-1970s. In the ICI approach, each guideword is combined with relevant process parameters and applied at each point (study node, process section, or operating step) in the process that is being examined.

Over the years, many organizations have modified the hazop analysis technique to suit their special needs (Ford and Brown, 1990). These various industry-, company-, or facility-specific approaches may be quite appropriate in the applications for which they are intended. For example, to guide teams more quickly to specific process safety areas, the original guidewords and the original process parameters have been modified and specialized lists of guidewords have been set up (Greenberg and Cramer, 1991). In order to investigate off-premises hazards, domino-effects-specific guidewords and parameters can thus be developed and used. Acikalin (2003) suggests that domino-accident escalation may take place due to three different effects: overpressure, radiation and missile projection. Considering domino effects, the literature (Fievez, 1996) also makes a distinction between 7 different accident scenarios. Based on these specific off-site data, guidewords and domino parameters can be listed as in Tables 6.3 and 6.4.

Domino hazop parameters focus attention upon a particular consequence aspect of the design intent towards a domino accident scenario. Domino hazop guidewords, when combined with domino hazop parameters, suggest possible

Table 6.3 Domino-hazop-analysis guidewords and meanings.

Domino guidewords	Meaning
1. Overpressure effects?	Possibility of overpressure effects inducing secondary (escalation) effects
2. Radiation effects?	Possibility of radiation effects inducing secondary (escalation) effects
3. Missile projection?	Possibility of missile projection inducing secondary (escalation) effects

Table 6.4 Domino-hazop-analysis parameters.

1. Boiling liquid expanding vapor explosion (BLEVE)	2. Vapor-cloud explosion (VCE)	3. Poolfire
4. Jetfire	5. Tankfire	6. Boilover
7. Explosion		

deviations on the process point under consideration possibly leading to a domino-effects scenario. The following example can illustrate investigating domino deviations using domino guidewords and domino parameters:

Domino hazop guideword		Domino hazop parameter		Deviation
Missile projection	+	BLEVE	=	Missile projection upon this process point might cause deviations (on this process point), possibly resulting in a (secondary) BLEVE.
Radiation effects	+	VCE	=	Radiation effects upon this process point might cause distortions (on this process point), possibly resulting in a (secondary) VCE.
Overpressure effects	+	Poolfire	=	Overpressure effects upon this process point might cause distortions (on this process point), possibly resulting in a (secondary) poolfire.

The suggested domino hazop analysis may be further elaborated and changed by the MPC to fit with multi-plant characteristics.

6.3 Summary and Conclusions

Over the last decade, a variety of computer-automated tools have been developed to determine the likely occurrence of domino effects. However, these tools do not offer transparent answers in terms of prioritizing the taking of domino-prevention measures in a complex multi-plant surrounding of chemical installations. Moreover, these tools only aim at taking precautions in terms of domino-safety risks, and fail to address domino security risks. Nevertheless, managerial decisions to prevent such catastrophic major accidents have to be made as efficiently and as effectively as possible. Decisions must be based not only on safety and security requirements but also on economic constraints.

For this purpose, a strategy should be used that allows domino-hazard analysts to compare the installations of an industrial area and determine a ranking with respect to the dangerousness of the installations both in terms of causing and continuing safety and security related domino effects. This ranking can then be used to take decisions concerning the prioritization of performing safety- and security-risk-analysis techniques in the area. Therefore, a methodology for the relative ranking of sequences of chemical installations in terms of their liability to produce escalating effects is proposed. The philosophy behind this approach is to address questions arising from the domino-risk analysis, thus determining the relative significance from a safety and a security point of view, before performing additional and much more costly large-scale hazard evaluation or risk-analysis studies.

To compare pairs of industrial installations in terms of their domino-danger contribution, quantitative data must be used. For this purpose, all different installation types must be defined. Consequently, these types must be linked to the possible major accident scenarios on the one hand and to scenario-failure-frequency data on the other hand. Using this information, a methodology was suggested focusing on the propagation aspect of an installation in a domino-accident sequence rather than on the intrinsic danger of the installation itself. A ranking list of installations contributing most to continuing domino events should be generated by a tool based on the suggested methodology.

The mathematical requirements and the working procedure for actually developing such domino computer program are given in this chapter. The approach for developing a computer-automated tool using simple installation information as input to first divide a multi-plant area into smaller areas, secondly to investigate these sub-areas for their scale-free behavior and third to rank domino installation sequences, should be elaborated for multi-plant decision makers (responsible for investigating complex areas in terms of complex security as well as safety-related domino effects). The use of such a method/tool would promote a more objective choice where prevention measures designed to alleviate catastrophic intentional as well as accidental escalation effects are concerned. Such a tool could prove particularly promising if installations in a multi-plant context have to be judged with respect to domino effects possibly affecting parts or all of the area under consideration and possibly affecting the area's environment.

7
Assessing, Evaluating and Continuously Optimizing Operational Staffing Levels Within a Multi-Plant Area

7.1
Introduction

A methodology for optimizing the quality and quantity of staffing levels in an industrial area should be designed and worked out.

Assisting chemical companies in evaluating and optimizing their operational staffing levels is essential in a multi-plant context. Two instruments in this regard are needed: an instrument for evaluating operational staffing levels as they currently exist within the industrial area (cluster, company, business unit, or installation), and an instrument to carry out a staffing-level-change risk analysis.

Furthermore, the manning-level resilience of a chemical plant or a chemical cluster and the potential to fulfill at all times all safety-critical tasks within a chemical industrial area might be of major importance to the developers of a multi-plant safety-management system. The size of operational teams in the chemical industry can have an important effect on the ability to control abnormal situations. For example, a lack of qualified operational personnel in unusual conditions and the resulting lack of process control can trigger a series of internal or external accidents, eventually leading to a multi-plant accident. Therefore, a practical method to evaluate the safety critical staffing levels required to meet performance specifications for safety critical activities is needed. For single plants as well as for clusters of chemical plants, the method should enable consultants and inspectors to consequently apply principles to assess those manning levels representing the last but one line of defense in the prevention of, among others, major and/or multi-plant accidents.

The importance of human factors, human errors, and human–machine interfaces for achieving high safety performance within a chemical plant is well documented in the academic and professional literature. Papazoglou *et al.* (2003) for example, present a QRA method whereby a technical model is linked with a management model to integrate the effects of the safety-management system into the quantification of risk of an installation handling dangerous substances. Bellamy *et al.* (2008) for example, describe a holistic model helping to understand the relationship between human factors, safety-management systems and other organizational issues within the chemical industry. Human factors literature,

Multi-Plant Safety and Security Management in the Chemical and Process Industries. G.L.L. Reniers

ISBN: 978-3-527-32551-1

although not *directly* explaining staffing-level assessment, is very important for *indirectly* understanding the underlying concepts, factors, indicators, models, systems, theories, *etc.* of staffing-level assessment in the chemical industry. The reader interested in the relationship between human factors and safety is referred to CCPS (1994a), Cacciabue (2004), HSE (2004), HSE (2005a), and CCPS (2007).

However, literature on the link between change management and operational staffing levels as well as performance management and operational staffing levels in the chemical industry remains very scarce.

Exploratory research by Zwetsloot *et al.* (2007) uses four case studies concerning staff reductions in the chemical process industry to build a conceptual model for optimizing operational shifts. Using people, management, technology, and safety related factors, the model suggested by Zwetsloot *et al.* facilitates an open discussion between all company stakeholders leading to so-called "optimum staffing arrangements" that are perceived as being acceptable *by all stakeholders*. However, the Zwetsloot model does not link staffing levels with performance management or quality management, nor does it allow an objective evaluation of existing operational manning levels within a chemical company to be carried out with the aim to continuously improve them.

Although manning levels are continuously evolving and changing within chemical companies, to date no systematic methodology for assessing the operational manning levels exists. In fact, at present, chemical organizations use data obtained by experience (trial-and-error approach) to decide on their operational manning levels. To streamline the manning-level change and assessment processes and to put forward a methodical and structured approach to enhance industrial practice on the subject, a benchmarking of best practices by highly experienced organizations can be used (Luu *et al.*, 2008). In this chapter, best available industrial practices and techniques on setting up, managing, controlling and optimizing operational manning levels were investigated, and expert opinions (of production, SESQ and HR managers of major chemical companies) were taken into account. A multi-disciplinary working group was formed composed of human-resources managers and health and safety managers from the participating multi-nationals.

Based on the multi-disciplinary experts' working group results, manning levels are defined as the filling in of functions within an organization. A function represents a combination of tasks that need to be carried out in an organization or a part thereof. A function thus depends on the competences of the person filling in the function (and carrying out the tasks). According to this definition, manning levels consist of two aspects: a quantitative aspect (i.e. a *number of functions*) and a qualitative aspect (i.e. the *competences of the functions*). A methodology for assessing and optimizing manning levels in an organizational context needs to take both the quantitative aspect and the qualitative aspect into account.

Furthermore, different types of staffing levels are defined:

a) Full staffing levels: Full staffing levels are defined as normal staffing levels multiplied by a factor that takes planned and expected absences from work into account;

b) Normal staffing levels: Staffing levels required to produce in an economic efficient way and taking all SESQ regulations and company guidelines into consideration;

c) Minimum staffing levels: Staffing levels required (i) to produce (during a predefined period of time) in an economic efficient way and taking all SESQ regulations and company guidelines into consideration (possible with a lower output, for example, tasks merely offering support can be postponed during a limited period of time) and (ii) to shut down an installation in an economic efficient and safe way;

d) Safety staffing levels: Staffing levels needed to handle a worst-case scenario and, if the chemical installation is still in operation, to shut it down safely.

The next section discusses the conceptual staffing-level-management model employed to elaborate the suggested staffing-level-assessment methodology.

7.2 Staffing-Level (SL)-Assessment Management Model

The model needs to take both the qualitative aspect and the quantitative aspect of staffing levels into account, and the user of the method (to be developed) should be able to assess and to evaluate management quality of both these SL aspects, irrespective of specific company characteristics. The SL optimization model is illustrated in Figure 7.1.

In-depth interviews with production, SESQ and HR managers from the participating companies revealed the theoretical existence of staffing-level demands and staffing-level resources. A major cause of errors is a mismatch between the staffing-level demands and the staffing-level resources. To guarantee effective and efficient organizational staffing levels, the difference between demands and

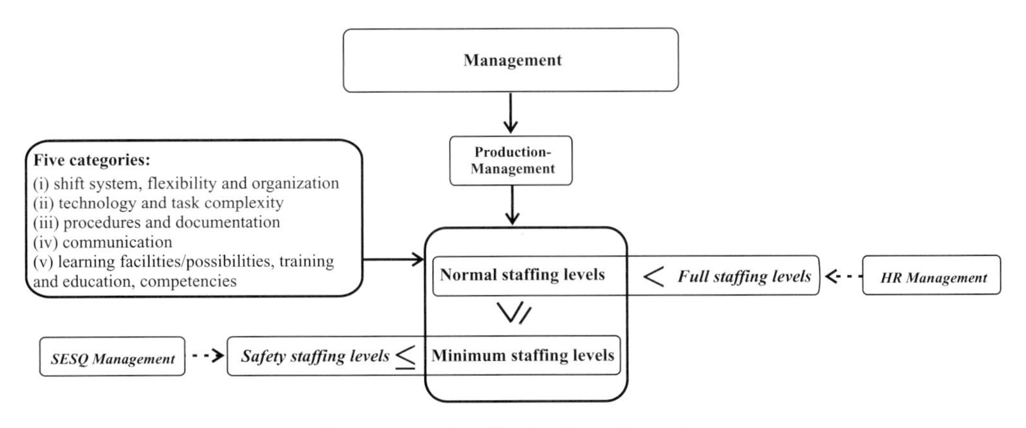

Figure 7.1 Management model for operational staffing levels in chemical organizations.

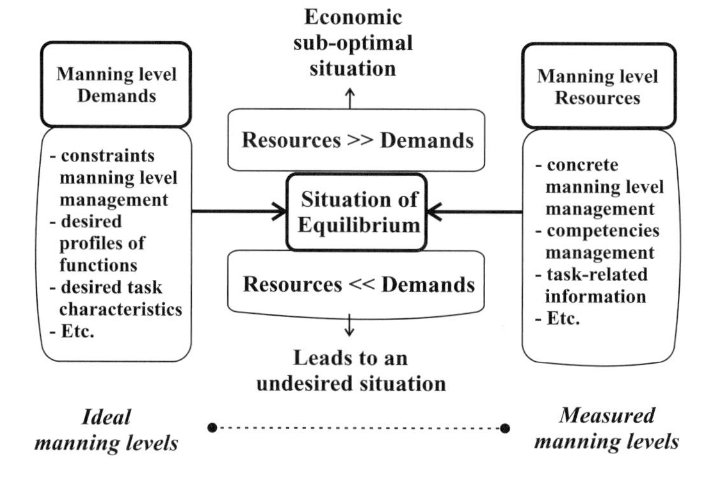

Figure 7.2 Manning-level demands *versus* manning-level resources in an organization.

resources need to be well considered and well chosen, and this needs to be accomplished for any circumstances and independent of installations or business units within the organization. Figure 7.2 illustrates the manning-level demands *versus* manning-level-resources concept.

Figure 7.2 shows that a distinction should be made between "ideal manning" representing the manning-level demands and "measured manning" representing the manning-level resources.

The former (ideal) manning concerns manning levels required by managers due to miscellaneous constraints, strategic reasons and/or management choices made in an organization. Manning-level demands (ideal staffing) concern ideally required personnel characteristics (e.g., ideal education, ideal on-the-job training, ideal working experience, ideal company theoretical training sessions, ideal mental qualities/features, *etc.*), task attributes (e.g., complexity, sequential order, time constraints, *etc.*), number of tasks that need to be carried out, organization of shifts, hardware and procedures concerning communication, *etc.*

The latter (measured) manning concerns manning levels which can be measured to certain extent by manning-level indicators. Manning-level resources (measured staffing) are determined by managing manning levels in general (how does communication take place within the organization, what are the working procedures at hand, *etc.*), as well as managing the qualitative part of manning levels (through competence profiles, learning facilities/possibilities, control of, for example, competencies, *etc.*) as well as managing the quantitative part of manning levels (via task schedules, the shift system, *etc.*).

The method to assess and to evaluate staffing levels in an industrial area should give recommendations about the way in which differences between ideal staffing levels and measured staffing levels are managed.

To systematize their evaluation, manning-level evaluation topics are divided into five categories (as can also be seen in Figure 7.1): (i) shift system, flexibility and organization, (ii) technology and task complexity, (iii) procedures and documentation, (iv) communication, and (v) learning facilities/possibilities, training and education, competencies. These categories are clarified in the existing staffing-level assessment (IESLA) instrument (provided in Appendix B). The approach and methodology behind the instrument is expounded in the next section.

7.3 Instrument for Existing Staffing-Level Assessment (IESLA)

The proposed instrument for assessing and evaluating existing staffing levels at an installation, a business unit or a chemical enterprise is based on systematically assessing the quality of actual management practices of staffing levels within the installation, business unit or chemical enterprise, thereby employing the five aforementioned categories.

For the instrument for existing staffing level assessment (IESLA) that is provided in Appendix B, the following textual design is used. Per category, the importance and the goal of the category as regards the operational manning levels are discussed. Next, via specifications of a series of general requirements, the manning level ideal demands in terms of the category being evaluated are given (indication for the *ideal manning*). Afterwards, by means of generic indicators the needs for manning-level measurements are identified (indication for the *measured manning*). Finally, several questions are provided in order to offer enterprises a simple checklist for comparing ideal staffing levels with existing staffing levels. As quality and quantity cannot be separated from each other when evaluating manning levels (cf. the definition of manning levels in Section 7.1), no explicit distinction is made here between these two aspects. Each question is followed by suggested (and well-considered) manning-level control measures enabling the IESLA user to efficiently and effectively carry out a self-assessment. It should be stressed that the IESLA instrument serves merely as a document giving a systematic *indication* to its user of the *quality of its operational staffing levels*. It certainly does not provide its user with concrete figures. A step-by-step company-specific procedure can then be drafted to help company, business unit or installation management take more concrete steps and actions to improve (if required) the (IESLA-pinpointed) manning-level categories for which improvement is desired.

Efficient and effective operational manning levels can be regarded as the adequate integration of three interrelated domains: job, human and organization (HSE, 2005b). Manning levels are concerned with what people have to perform within the organization (tasks and task characteristics, the "job"), with whomever carries out the tasks (the individual, the shift and their (personal and aggregated) competencies, the "human"), and with the environment in which the job has to be filled in (the company and its characteristics and culture, the "organization"). Moreover, manning levels are influenced by a wider social context, local as well as national.

The domain job includes all subjects related to the job *in se* (e.g., tasks, documentation, workload, the working environment, displays, controls, *etc.*). The

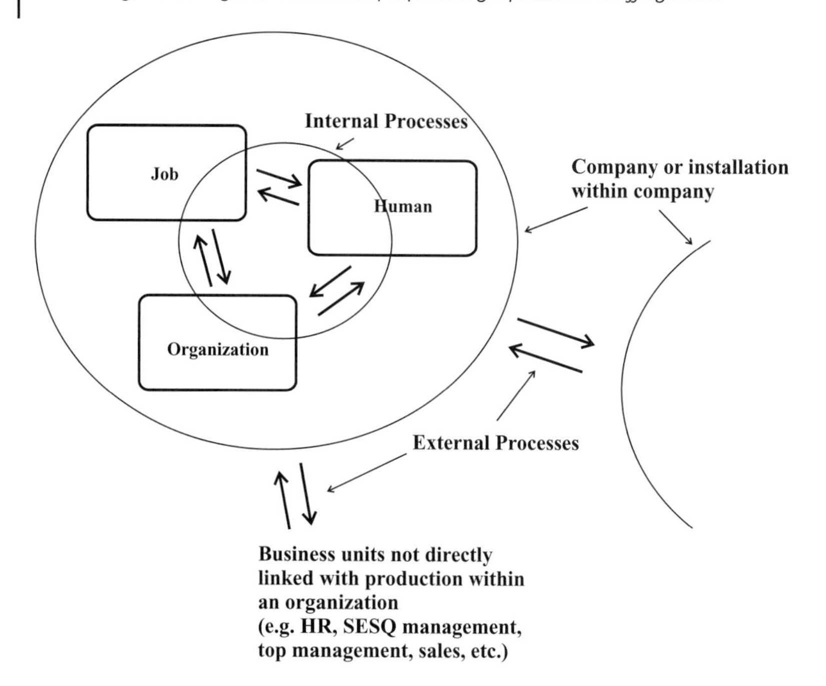

Figure 7.3 Integration of domains determining the manning-level quality in a chemical industrial area.

domain human involves all inter-relational and person-related characteristics (e.g., competencies, personality, skills, attitudes, risk perception, social context, individual and group characteristics, *etc.*). The domain organization is concerned with management processes and management practices, including the required internal processes that design and mold the integration, the elaboration and the inter-dependence of the three domains, as well as the required external processes that shape and form the integration, the elaboration and the inter-dependence of the three domains with the external environment (e.g., the culture of the workplace, leadership, resources, communications, work patterns, *etc.*). Figure 7.3 illustrates the manning-level domains and their inter-relationship.

Depending on the answer given to a certain question in the IESLA, improvement (in terms of operational staffing levels within the installation, business unit or company) might be indicated or desired. The manager responsible for optimizing the staffing levels concerning this question needs to investigate within which domain(s) these improvements can be realized. To this end, Table 7.1 may serve as an expert-based aid. The table was filled in and discussed by a team of 6 experts from 6 different enterprises. This suggested table is only guidance and may be filled in and completed at company level (e.g., by company experts).

For example, if an answer in response to Question 3 regarding the category "communication" indicates a potential advancement of staffing-level quality within the company, business unit or installation, Table 7.1 shows that in this respect

Table 7.1 IESLA table of improvement (domain classification per question for improving the existing staffing levels).

Question	Shift system, flexibility and organization	Technology and task complexity	Procedures and documentation	Communication	Learning facilities/ possibilities, training and education, competencies
	Domain classification				
1	H, O	H, O	O	H, O	H, J, O
2	H, O	H, O	H, O	H, O	H, O
3	H, J, O	J, O	H, O	H, O	H, J, O
4		O	H, O		O
5		O			H, O
6		H, O			
7		H, O			

Key: H = Human, J = Job, O = Organization.

(concerning this particular question) the domains human and organization should be tackled.

The IESLA is drafted as an easy-to-use self-evaluation tool (see Appendix B). The detailed checklist of clear and unambiguous questions can be used directly by enterprise personnel without the need for any further requirements. The current quality of managing the staffing levels is thus easily evaluated.

For further improving the manning levels, measures should be taken in each domain, based on the results of implementing the IESLA checklist and its user-specific Table 7.1.

The IESLA instrument should not be used for large or substantial manning-level changes. A method to tackle such large SL changes is elaborated in the next section.

7.4 The MCSL Method

Two types of staffing-level changes are possible: small (latent, daily) changes and large (ostentatious, exceptional) changes. The former changes mainly depend on task rescheduling due to, for example, automation or small installation changes and can be termed as "existing staffing levels". The latter changes depend on, for example, substantial changes on chemical installations or placing a new installation on the site. They can also result from various exceptional circumstances such as wild strikes, epidemics, security issues, *etc.* The suggested methodology to evaluate changes of existing staffing levels (MCSL) uses a staffing-level risk matrix to assess the risks accompanying (important) planned or unexpected (theoretical scenario-based) manning-level changes. Hence, the MCSL tool is not used to analyze and evaluate day-to-day changes in existing staffing levels (for which the IESLA instrument can be employed, see previous section), but is designed to deal with substantial staffing-level changes.

Basically, two possible large staffing-level changes within a multi-plant area, company, installation, or business unit are possible:

Case (i) manning-level requirements (= manning-level demands) are altered (while manning-level resources initially remain constant);
Case (ii) manning-level resources are changed (while manning-level demands initially are unchanged).

The first case may be the result of a management decision on reducing or increasing the staffing levels because several tasks have been ceased or recently introduced within the installation/business unit/company. In this case, manning-level demands are adjusted, and as a result, the manning resources need to be modified as well. The change in the resources is then driven by newly defined (long-term) target staffing levels.

The second case may occur, for example, with the outbreak of a heavy epidemic or a pandemic. Due to such a possible disaster, manning-level resources might change dramatically. Hence, although manning-level requirements have not undergone a transformation there is a short-term abrupt altering in the manning-level resources.

In both cases, a risk analysis should be performed depending on the underlying scenario. In the first case, the modified manning-level demand conditions define the supply, while in the second case, the shifted manning-level supply conditions give direction to the manning-level demands. The purpose of the risk analysis should be to diagnose the difference between demand and supply in an objective way. A risk-analysis methodology for assessing possible large staffing-level changes is therefore elaborated hereafter.

Staffing-level changes used in the context of this book are defined as changes in manning levels with the potential of a serious impact on production and/or SESQ within a multi-plant area, a company, an installation or business unit. This can be a quantitative change (alteration in the number of functions), a qualitative change (alteration in the required competencies for a certain function, such as changes in tasks, job contents and necessary qualifications), or it might be the combination of both quantitative and qualitative changes. These manning-level changes might take place in various manners: expected/unexpected, voluntary/involuntary, internal/external, gradually/abruptly, *etc.*

As in every risk analysis, responsible management (in the multi-plant area the MPC data administration is responsible) should start by deciding on the responsibilities, team composition, objectives, working method (description), necessary information, *etc.* In both the above cases, a manning-level supply (resources) change is considered to be the reference for the risk analysis. In the first case, a translation from the altered target staffing levels to a certain (qualitative and/or quantitative) increase or decrease of employees available for a specified function should initially be made. In the second case, a specific increase or decrease (quantitative and/or qualitative) of employees available for a specified function is assumed per scenario. With "an increase or decrease of employees available for a specified function" one of the following three possibilities below is implied:

a) *Quality* of employees within a certain function changes (e.g., an employee with function *A* has had extra training for a certain task or an employee with function *B* has not passed in an evaluation for carrying out certain actions of standard emergency procedures);

b) *Quantity* of employees within a certain function changes (e.g., the number of employees with function *A* increases from *x* to *y* or decreases from *r* to *t*);

c) A combination of quantity and quality changes of employees within a certain function.

As already mentioned, the starting point for carrying out our "manning-level change risk analysis" is a change of the manning-level resources. A comprehensive, complete and clear description of this change is used to determine the resulting qualitative and/or quantitative alterations of staffing levels.

Company experts (e.g., production managers, shift leaders, HR managers, SESQ managers, a combination of the previous, *etc.*) need to assess the impact of manning-level changes on production, quality, health, safety, security and environment. It should be noted that it is possible for the experts to decide that the adding or omitting of one person with specific qualifications within a certain function has a larger impact on production and/or SESQ than, for example, three other persons within the same function but with different qualifications.

Using a decision matrix (see Figure 7.4), the severity of the impact of manning-level modifications on production and/or SESQ are taken into consideration (there

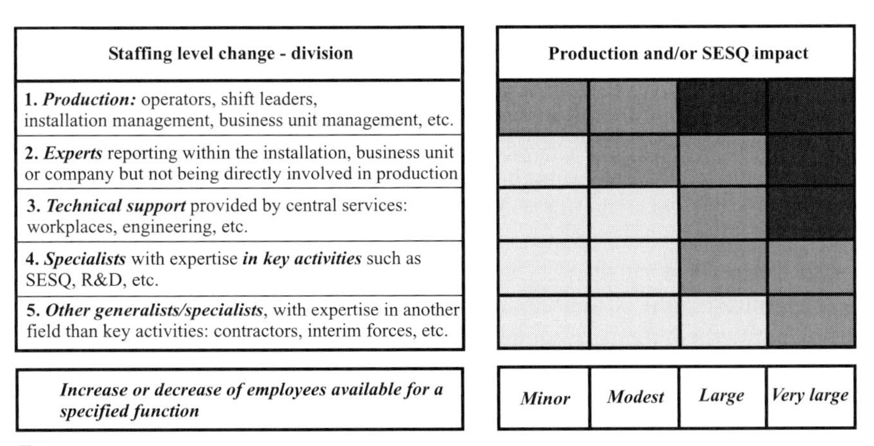

Figure 7.4 Decision matrix for determining the team of a staffing-level-change risk analysis.

are three possibilities: limited consequences, significant impact, and large impact. Completing the table allows the composition of the team entrusted with performing the actual manning-level risk analysis to be fixed. Although it is recommended that either the head of the SESQ department or the head of production (or a combination of the aforementioned) completes the table, the decision about which function(s) actually complete(s) the decision matrix from Figure 7.4 is to be made within the MPC or within each individual company.

For each of the possibilities (10%, 40% or 70% gray) the team to carry out the manning-level-change risk analysis is assigned (only functions are indicated, no names). The procedure takes into account the specified divisions that suffer from an impact resulting from a change in their staffing levels.

A short theoretical example to illustrate the above working procedure is given hereafter. According to the experts, staffing-level changes being considered for a particular function in division 1 (production) have, for example, a moderate impact on production and/or SESQ and staffing-level changes in division 3 (technical support) have a very large impact on production and/or SESQ. The decision matrix then shows that the production and/or SESQ-impact of proposed manning-level changes imply a 40% gray impact and a 70% gray impact respectively. The organization subsequently has to compose a team (only functions, no names) (different gray percentages give rise to different pre-defined team compositions) to perform the actual manning-level change risk analysis. When different divisions give rise to different team compositions, the rule is to form a team as complete as possible (functions are rather added to, than subtracted from, the risk-analysis team). Organizational procedures could for example call for a team consisting of the company manager, the shift leader, the head of SESQ department, the director of central support and the person responsible for maintenance at an installation.

The experts investigate the potential impact of the suggested manning-level changes by comparing the new situation (with changed manning levels) with the current situation (with current manning levels) as regards production and SESQ. For this purpose, a checklist is used (see Table 7.2). Every item for which the experts consider the manning-level changes impact to be relevant, is examined and analyzed.

In case (i) (see the beginning of Section 7.4), additional information about the expected changes of manning-level resources on the item of relevance should be provided. A list of action points to absorb – whenever and wherever necessary – the negative impact of the changes is put together. Finally, the new manning-level resources (taking the action points into account) are evaluated *vis-à-vis* the (management-imposed) new manning-level demands.

In case (ii) (see the beginning of Section 7.4), additional information about the expected changes of manning-level resources on the item of relevance should be provided as well. An explanation needs to be provided about the (indirect) abrupt manning-level resources changes occurring for the item under consideration, resulting from the scenario. A list of (temporary) action points to absorb – whenever and wherever necessary – the negative impact of the changes is put together, and the timing constraints of the suggested action points are clearly given. Finally, the new manning-level resources (taking the action points into

Table 7.2 Checklist for evaluating the impact of manning-level changes on production and SESQ (non-exhaustive).

Production and SESQ checklist		*Relevant impact?*		
IMPACT ON (non-exhaustive)	**Possible items of concern (non-exhaustive)**	***Job***	***Human***	***Organiz.***
1. Shift system, flexibility and organization				
1.1. Functions	Filling in of functions	☐	☐	☐
1.2. Responsibilities	Workload distribution	☐	☐	☐
1.3. Housekeeping	Control of housekeeping	☐	☐	☐
1.4. First aid	Availability of first aid	☐	☐	☐
1.5. Management/delegate	Management directives	☐	☐	☐
1.6. Redistribution and execution of tasks and actions	Shift flexibility	☐	☐	☐
1.7. Flexibility	Flexibility within business unit, installation and/or company	☐	☐	☐
1.8. Other factors		☐	☐	☐
2. Technology and task complexity				
2.1. Employee workload	Complexity, volume, impact on remaining employees	☐	☐	☐
2.2. Personal protective measures		☐	☐	☐
2.3. Material maintenance		☐	☐	☐
2.4. Material controls		☐	☐	☐
2.5. personnel qualifications		☐	☐	☐
2.6. Various circumstances	Various situations or accident scenarios	☐	☐	☐
2.7. Human errors		☐	☐	☐
2.8. Other factors		☐	☐	☐
3. Procedures and documentation				
3.1. Working method	Instructions	☐	☐	☐
3.2. Working procedure	Requirements for safely carrying out the job	☐	☐	☐
3.3. Various circumstances	Various situations or accident scenarios	☐	☐	☐
3.4. Emergency planning	Impact on emergency plan	☐	☐	☐

Table 7.2 *Continued*

Production and SESQ checklist		***Relevant impact?***		
IMPACT ON (non-exhaustive)	**Possible items of concern (non-exhaustive)**	***Job***	***Human***	***Organiz.***
3.5. Legislation and regulations	Impact on the capability required to comply with legal requirements	☐	☐	☐
3.6. Company directives	Impact on the capability required to comply with company requirements	☐	☐	☐
3.7. Other factors		☐	☐	☐
4. Communication				
4.1. Communication within business unit or installation		☐	☐	☐
4.2. Communication within company		☐	☐	☐
4.3. Various circumstances		☐	☐	☐
4.4. External communication		☐	☐	☐
4.5. Other factors		☐	☐	☐
5. Learning facilities/possibilities, training and education				
5.1. Competences	Remaining personnel, lost competences	☐	☐	☐
5.2. Training and education	Requirements for remaining employees	☐	☐	☐
5.3. Learning trajectory		☐	☐	☐
5.4. Other factors		☐	☐	☐

account) are evaluated *vis-à-vis* the (management-imposed) new manning-level demands for the indicated time period.

Although Table 7.2 is non-exhaustive, the different items of consideration mentioned for each of the five categories in the table should at least be included in the checklist. Other items may be added, depending on the specific needs of the company.

For every item from Table 7.2 where a relevant impact of the suggested manning-level changes is expected by the experts, the latter changes are linked to the manning-level resources and a thorough analysis is made (the relationship is explained). If the deviations resulting from the impact are considered to be negative and unacceptable, action points have to be put forward. In case (i), action points should be permanent, while in case (ii) a certain duration (period of time) has to be determined for every action point to indicate how long these efforts can be sustained.

To verify the effectiveness of the suggested action points towards eventual (major) problems, a checklist to control the effectiveness of the points of action to absorb negative impacts of the manning-level changes in the case of potential major problems, can be drafted. An assessment and evaluation should be made of (in case of problems) covering

a) supervision and intervention possibilities (problem detection);
b) distraction possibilities (problem detection);
c) obtaining correct information (problem diagnosis);
d) communication links (problem diagnosis);
e) manning assistance in case of problems (problem recovery);
f) recovery operations (problem recovery).

To help the user with this assessment, an evaluation instrument (provided in Appendix C) is elaborated in the form of a checklist to consider safety-critical tasks in the control room as well as in the field. The method is developed to cope with different possible physical configurations within the staffing-level arrangements in an industrial area. The possible configurations are:

a) One or more operators with all-round competence are responsible for the activities in the control room (CR) and in the field;

b) One or more CR operators who permanently staff the control room and one or more field operators working only in the field.[1] Eventually, within both groups, competences can be arranged in such a way that operators are interchangeable;

c) Field operators who are not permanently present, but who may be summoned from elsewhere within the plant/cluster to assist one or more permanent CR operators under safety-critical situations;

d) Control-room operators and field operators not constantly present, but who may be summoned from elsewhere within the plant/cluster to ensure a safe situation under safety-critical circumstances.

During the staffing-level assessment, the safety and the lay-out of the industrial area is subject to evaluation. The lay-out of the multi-plant area and/or the plant and of the control room is investigated in terms of whether operators are able to move from one point to another within certain time limits, the consequences resulting when operators are not able to be in a particular place within a given time are identified and the reliability of the supporting equipment and the supporting documentation is questioned.

In summary, the evaluation verifies whether safety-critical staffing levels in an industrial area affect the reliability and timeliness of detecting safety-critical

1) With the term "field", the physical area constituting the installation and its immediate surroundings is meant.

Table 7.3 Parameters and principles to deal with occurring operational staffing-level problems.

Operational SL problem parameter	Principle	Interpretation
Operational SL problem detection	1. Supervision/ intervention possibility	There should be continuous supervision of the process by skilled field and CR operators and the possibility to intervene whenever needed.
	2. Distractions	Distractions such as answering the phone, talking to people, performing administrative tasks and acting upon nuisance alarms should be minimized to reduce the possibility of missing/overseeing/responding too late to alarms.
Operational SL problem diagnosis	3. Information	Sufficient information required for diagnosis and recovery should be easily accessible, correct and intelligible.
	4. Communication links	Communication links at single-plant level as well as at cluster level between the control room and the field as well as between different control rooms should be reliable.
Operational SL problem recovery	5. Assisting personnel	Staff required for assisting in diagnosis and recovery should be available in time and with sufficient time to attend when required.
	6. Recovery operations	Operating staff should be allowed to concentrate on recovering the plant to a safe state. Necessary but time-consuming activities should therefore be allocated to others, for example, summoning emergency services or communicating with adjacent plant-safety management. Moreover, all recovery operations should be executed on time.

Source: Reniers *et al.*(2007a).

problems, diagnosing them, and lifting recovery to a safe state. Therefore, conditions such as the number of people required, the means necessary and the competences needed to be able to guarantee safety in the case of calamity is checked against six principles (see Table 7.3).

The instrument is conceived to identify the possible bottle-necks of personnel organization and to find a solution for a problem under abnormal circumstances. To do this, four questions for every principle are addressed:

a) Can the principle be violated?
b) In what way has the principle been/could the principle be violated?
c) What are the action points to counter the violation?
d) Are these action points reliable and effective?

Failing on one of the above questions indicates a gap in the industrial area's personnel organization and implies that actions are required to deal with this failure. Because each "failure" is different, a gradation is added to the different assessments, for example, A is considered "best practice", D "worst practice". If an industrial area fails to be situated in the A or B category, the staffing-level safety in the area is insufficient and some measures are needed. Directives for such measures are given in the evaluation instrument itself. For example, to be able to improve from category C to category B, a back-up alarm-warning system might be installed. Table 5.9 offers an easy-to-use ranking tabulation, indicating the company safety-critical staffing-level failures.

The implications of the ranking degrees A, B, C or D in the case of the safety critical SL instrument in terms of the final decision are given in Table 7.4.

To enhance the objectivity of staffing-level assessment and evaluation, certain documents are required to verify the answers when filling in the checklist. When

Table 7.4 Description of the different gradings for the safety-critical-staffing evaluation instrument.

Degree	Description
A	The parameter of the safety critical problems is guaranteed by the inherent presence of a sufficient quality and quantity of staffing levels. The organization of the industrial area personnel guarantees safety in safety-critical circumstances.
B	The parameter of the problem is guaranteed. There is no need for any kind of back-up system to solve the safety critical problem. However, there is one disadvantage: the quality and/or the quantity of the staffing level is not sufficient to guarantee the inherent presence of competent personnel, information and/or communication in the case of safety-critical circumstances. Nonetheless, the problem can be tackled promptly addressing the qualitative and/or quantitative staffing levels required.
C	The quality and/or the quantity of the staffing levels suffice to solve safety-critical situations thanks to the presence of back-up systems. Measures are needed to ameliorate the response rate with which the staffing levels are recovered and/or to ameliorate the quality and/or the quantity of the staffing levels.
D	The organization of personnel fails. For this problem parameter, the staffing levels should ameliorate in a qualitative and/or quantitative manner. The industrial area is not capable of guaranteeing safety and preventing incidents in the case of abnormal circumstances.

evaluating an answer from the questionnaire, the final judgment depends on the evidence provided for answering the checklist. If the answers accompanied by the necessary documents are considered insufficient to underwrite staffing levels quality and/or quantity *vis-à-vis* SESQ and/or production requirements on a specific topic, the area fails for this topic. Extra documents that might be needed include estimation calculations or experiments (simulations) regarding the amount of time needed to react to incidents, data of previous accidents and/or observations of exercises, reliability studies of critical equipment, *etc.*

7.5 Roadmap of Staffing-Level Assessment

Staffing-level assessments must be planned and monitored just as any other process. The planning aspect requires ways to identify the specific initiatives to be taken on, while monitoring requires methods for measuring and assessing staffing level progress. To this end, a user-friendly staffing-level roadmap of assessment is elaborated. Since organizations differ in their activities, cultures, sizes, *etc.*, the elaboration of such a roadmap is not at all an easy task. As a pre-requisite for the roadmap, companies and their top management should be convinced of the importance of following-up their operational staffing levels.

This section discusses the design of a staffing level management of change process, that is, a series of consecutive steps, to guarantee that staffing-level changes (whatever they may be) are properly assessed and that actions taken due to the staffing-level changes lead to an acceptable situation (as regards production and SESQ) within an industrial area, an enterprise, installation or business unit. On the one hand, implementing the suggested roadmap may require evolution in a company's culture. It may also demand significant commitment from line management, departmental support organizations and employees. On the other hand, applying the principles and techniques of the management of change approach for assessing operational staffing levels described in this chapter may lead to substantially more efficient staffing levels and may be financially highly beneficial.

The suggested roadmap for assessing manning levels within a chemical industrial area is illustrated in Figure 7.5.

Since the MCSL method expounded in Section 7.4 can only be employed to assess the risks accompanying large manning-level changes, in the first stage of the roadmap, its user has to identify whether the staffing-level changes are small or large.

In the case where the staffing-level changes are identified to be small, the instrument for existing staffing-level assessment (IESLA, see Section 7.3 and Appendix B) can be used to analyze and to evaluate the staffing levels. The suggested IESLA uses a list of questions and measures to assess the staffing levels in a simple, unambiguous and objective way within a multi-plant area or a part thereof. In this way, self-evaluation and improvement (if required) of the staffing levels currently present within the area under investigation is possible.

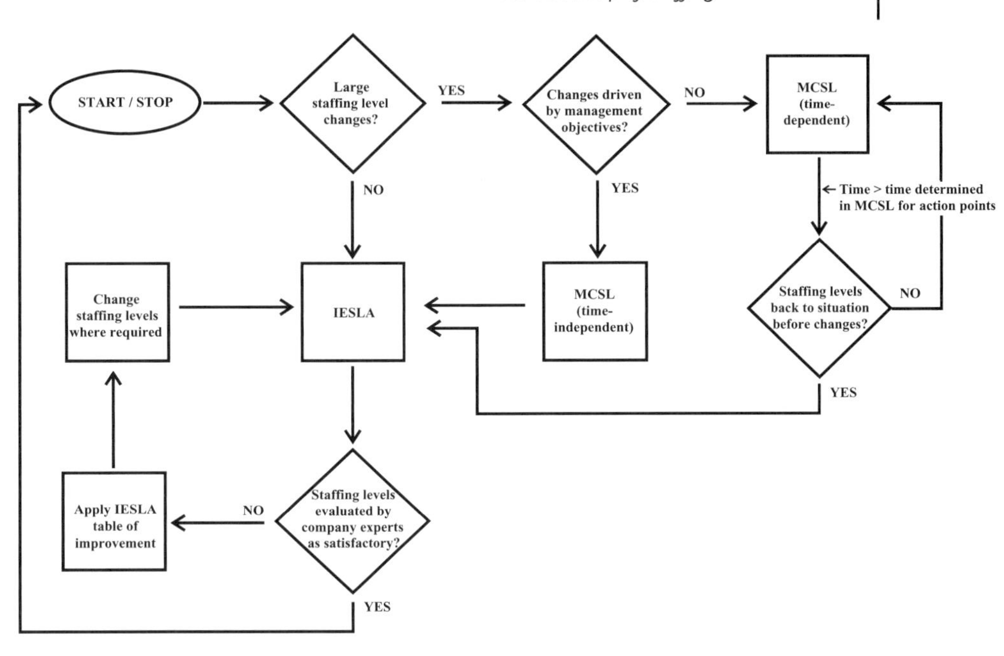

Figure 7.5 Roadmap of staffing-level assessment.

In the case where the staffing-level changes are termed large, the question whether the changes are time dependent or not needs to be answered. Due to the difference in time dependency of the action points between two cases – in the first case, action points should be permanent, while in the second case a certain duration (period of time) has to be determined for every action point to indicate how long these efforts can be sustained (see Section 7.4) – the MCSL should be applied in a different way in the roadmap of staffing-level assessment.

Once experts have decided on the action points after using the MCSL, the next step is to carry out the IESLA taking these action points into account. In this way, the new staffing-level situation may be verified and the staffing levels can be further optimized.

The suggested roadmap for staffing-level assessment allows taking both small staffing-level changes and large staffing-level changes into consideration, and serves as an easy-to-use tool to establish optimal staffing levels within any area of chemical installations (multi-plant (in the case of time-dependent MCSL applications), single plant, installation, factory, business unit). As a pre-condition to the proposed roadmap, the MPC and/or plant management should establish whether there is a procedure for its periodic application and if it is effective. A yearly interval can be recommended for the roadmap's application. The roadmap should also be employed in the case of special circumstances where large staffing-level changes are desired by management (time independent) or may be expected (time-dependent) or in the case of simulation exercises. The approach might, for example, be used by the MPC data administration to carry out staffing-level simulative exercises to be prepared for a pandemic outbreak.

7.6
The Way Towards Continuous Staffing-Level Improvement in Industrial Areas

As mentioned in Section 7.1, for optimizing the operational manning levels within a chemical company, there should be an adequate match between ideal manning levels and measured manning levels at all times and under various circumstances. A procedure can be developed by the multi-plant council and/or plant management to continuously monitor the manning levels and to optimize them if possible and/or to change them if needed.

Technology, automation, installation and machine safety and security, human resources management, *etc.*, depend on company characteristics and are thus highly company specific. Hence, by establishing the manning levels every organization needs to draft its own company-specific manning-level standards and, by using these standards, optimize their operational manning levels. To help companies in this staffing-level standards design process, the MPC may provide benchmarking results and expertise.

Four types of manning-level standards should be developed: management manning-level indicators, management manning-level objectives, concrete manning-level indicators, and concrete manning-level objectives. For more information on developing such performance standards, the interested reader is for example, referred to Rampersad (2004), Parmenter (2007), or Smith (2007). The relationship between the well-known Deming cycle of quality management and the four different types of manning level standards is shown in Figure 7.6.

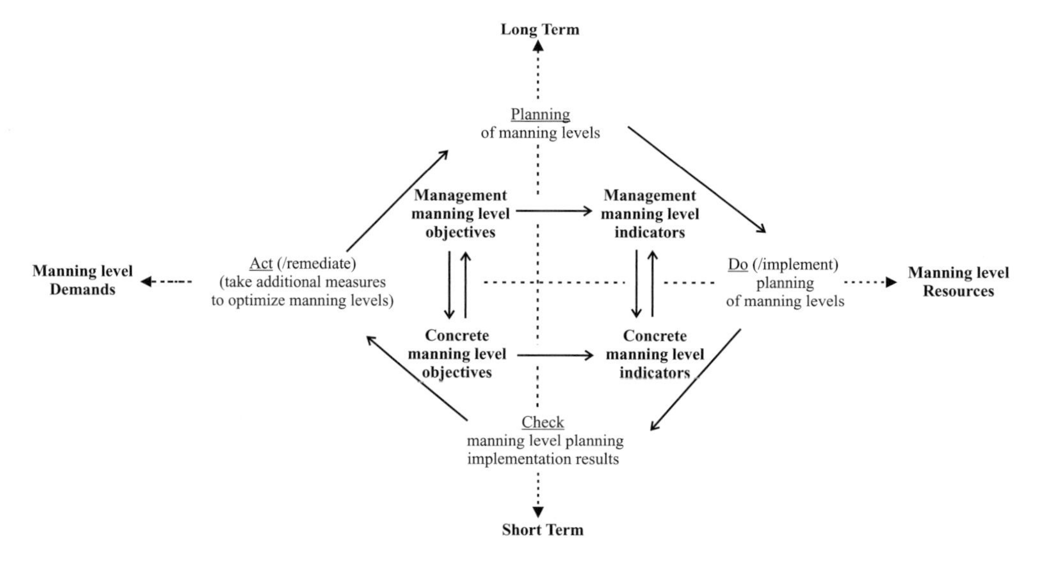

Figure 7.6 Overview of the relationship between SL standards and the PDCA-cycle.

Two fields of tension are displayed in Figure 7.6: on the one hand manning-level demands *versus* manning-level resources, and on the other hand long term *versus* short term. For example, the short-term manning-level resources are situated between the do-phase and the check-phase of the Deming cycle and can be steered by using concrete manning-level indicators. All other manning-level standards can be viewed in the same way.

Management manning-level indicators are defined as factors/indicators that are crucial for ensuring a chemical organization's manning-level continuity in the long term. Concrete manning-level indicators are measured quantities or qualities providing an adequate indication of management manning-level indicators' successes or failures. They can also provide an idea of the success or failure of the functioning of a process in the short term. Hence, using concrete manning-level indicators should lead to an accurate up-to-date overview of an organization's short-term manning-level resources. The measurements provide "realistic manning-level scores" and are essential for informing company or business unit management about all kinds of (small) daily manning-level changes.

Management manning-level objectives and concrete manning-level objectives, respectively, are quantitative figures or qualitative figures set to be achieved by company management or by business-unit management for a specific management manning-level indicator and a specific concrete manning-level indicator, respectively. Management manning-level objectives (e.g., with a tri-monthly or yearly frequency) and concrete manning-level objectives (e.g., with a weekly or monthly frequency) are used to control and remediate the organization's manning levels. Values of manning-level figures/qualities objectives result for example, from strategic considerations, company past experiences, benchmarking, *etc.* To develop the manning-level indicators, all influencing factors on manning-level qualities and quantities within an organization are to be determined. These factors might be functions, technology, automation, tasks, training, safety measures, management agreements, *etc.*

The three manning-level domains (job, human, organization) are used to categorize the influencing factors. Ideally, a maximum of 15 management indicators are employed and several tens of concrete indicators are used. Per domain, management indicators are determined. Per management indicator, several concrete manning-level indicators are chosen. Table 7.5 shows a number of management manning-level indicators examples.

Management manning-level objectives (which should set the targets for the management manning-level indicators of Table 7.5) might be for example: to carry out a document audit twice a year; to have shift and individual job evaluation conversations yearly; to have a 360° feedback evaluation twice a year; to verify the efficiency of internal management processes yearly; *etc.* These targets then result in an objective evaluation of the indicators of Table 7.5, which in turn leads to the indicators' changes, elucidations, completions, concretizing, *etc.*

Figures (and an objective notion) on the management manning-level objectives are obtained via the evaluation of concrete manning-level objectives. The latter objectives set the targets for the concrete manning-level indicators. Such concrete manning-level indicators should be specific, measurable, acceptable, relevant, and

Table 7.5 Non-exhaustive list of management manning-level indicators per manning-level domain (illustrative example).

Domain	Job	Human	Organization
Management manning-level indicators	1. Documents comply with all (legislative and company-related) requirements. 2. Documentation is correct, complete, and up-to-date. 3. Documentation is drafted from a process point-of-view (re-active as well as pro-active). 4. The complexity of task demands is adjusted to the available human factors. 5. The required competencies for carrying out all required tasks within the installation, business unit, or company are present. 6. *Etc.*	1. Every person is aware of all necessary technical and meta-technical knowledge and know-how for carrying out his/her function. 2. Every person is part of a social network/entity within a business unit or company. 3. Every person fully understands his/her responsibilities and acts as such. 4. *Etc.*	1. Management is transparent. 2. In the management process (independent of its organizational level) normal (standard) as well as various (special) circumstances/situations are taken into account. 3. The management process considers re-active as well as pro-active elements and follows the PDCA cycle. 4. The adequate functioning of internal processes within an installation or business unit is part of the company culture. 5. Internal processes support the manning-level domains and facilitate the interactions between them. 6. The adequate functioning of external processes within an installation or business unit is part of the company culture. 7. External processes facilitate the interactions between business units or companies and other business units or companies. 8. *Etc.*

traceable. Elaborating a concrete manning-level indicator matrix can for example, be achieved by drafting a questionnaire to be filled in by (production, SESQ, HR) company or business unit management. Table 7.6 shows an example of such a matrix.

After companies have drafted a company-specific concrete manning-level indicators matrix, they have to determine company-specific concrete manning-level

Table 7.6 Non-exhaustive list of concrete manning-level indicators per manning level category and per manning-level domain (illustrative example).

Categories	Domains		
	1. Job	**2. Human**	**3. Organization**
1. Shift system, flexibility and organization	– Be able to correctly carry out action X within time Y – Percentage of incidents in which tasks could not be carried out or were not carried out according to prescriptions – Percentage of incidents in which tasks could not be carried out or were not carried out according to prescriptions in simulation exercises of accident scenario X – Number of functions within a shift – Interchangeability between shift X and shift Y of a business unit or company – Multi-functionality of tasks – Extent to which the physical workload can be maintained within a shift – Extent to which supporting personnel (of another installation, business unit or company) can be employed (required time, required know-how, *etc.*) – (If applicable) time required for physical displacement between control room and business unit/installation	– Satisfaction scores within a shift per period X – Scores of involvement and appreciation in received feedback – Labor productivity of shifts – Percentage employees praising their job as being challenging – Average time period that a shift member holds the same position within shift – Percentage sickness/absenteeism per shift in period X – Percentage sickness absenteeism within installation or business unit per period X – Percentage staff turnover per shift per period X – Percentage staff turnover within installation or business unit per period X – Percentage operators where personal ambitions coincide with business unit or company ambitions	– Percentage available competences per shift per period X – Satisfaction scores within an installation or business unit per period X – Satisfaction scores within the company per period X – Number of meetings within installation or business unit to discuss manning levels – Number of meetings within company to discuss manning levels – Percentage sickness absenteeism within company per period X – Percentage staff turnover within company per period X – Number of functions within installation or business unit – Level of inter-changeability of human factors within company – Number of rounds of HSE personnel per period per business unit – Number of different disciplines involved in carrying out human factor risk analyses

Table 7.7 *Continued*

Categories	Domains		
	1. Job	**2. Human**	**3. Organization**
2. Technology and task complexity	– Required time to obtain all necessary information in case of emergency scenario X for an installation of complexity Y – Required time to handle accident scenario X for an installation with complexity Y – Required number of operators with competence X to handle accident scenario Y – Time required to carry out a task with characteristic X (e.g., composed of a number of sub-tasks) by function Y – Number of tasks within function X – Technological complexity of an installation or a business unit – Number of tasks that need to be carried out per person	– Time required for function X to carry out the correct initial handlings in case of accident scenario Y – Recovery time for accident scenario X needed by person with competence Y or by shift with competences R, S, *etc.* – Level of complexity of tasks per operator to be carried out – Number of tasks to be carried out per operator in special (non-standard) circumstances – Level of complexity of tasks per operator to be carried out in special (non-standard) circumstances – Level of complexity of an installation, a business unit and the environment and the impact on the human factor – To be able to safely carry out action X – Level to which responsibilities are well known – Satisfaction scores w.r.t. user expectations of technology – Level to which displays offer an overview of the process for the operator, together with essential information of individual processes	– Degree to which electronic safety systems are present – Level of automation as regards control on human failures – Reliability of control systems as regards human errors – Level of automation as regards responding to unexpected and undesired process conditions – Criteria for the adequate adaptation of task lists – Quality of the evaluation process of drafting working procedures – Quality of the instruction process of drafting working procedures – Level to which non-standard scenarios are included in management guidelines in combination with the use of technology – Number of meetings per time period as regards the effectiveness, the efficiency, the user-friendliness, *etc.*, of documentation, user manuals, working procedures, *etc.* – Number of meetings per time period as regards the complexity of the relationship between human/competences and technology/machines – Level of agreement between business unit procedures and company procedures – Number of human-factor analyses (job analyses, Kinney and Fine, *etc.*) carried out per time period for the human/technology interface – Degree to which emergency shutdown modes are present

3. Procedures and documentation	– Degree to which working procedures of an installation or a business unit are complete (and do not contain superfluous information) – Level of adaptability of installation or business unit working procedures – Level of availability of user manuals concerning installation X – Level of availability of electronic documents – Level of availability of hardcopy documents – Level of standardization of documents – Level of user friendliness of procedures – Degree to which procedures take non-standard circumstances into account	– Degree to which working procedures of a business unit or installation are easy to understand – Degree to which user manuals are understandable concerning installation X – Level of standardization of information exchange within and between shifts – Degree to which procedures take non-standard behavior into account	– Level of completeness of company-wide procedures and directives – Degree to which company-wide procedures and directives are easy to understand – Degree to which company-wide procedures and directives are easy to adapt – Degree to which business unit or installation layout takes accesses, freedom of movement and overview into account

Table 7.7 *Continued*

Categories	Domains		
	1. Job	**2. Human**	**3. Organization**
4. Communication	– Availability of communication within installation or business unit infrastructure – Percentage of communication errors within shift – Percentage of communication errors within business unit or installation – Number of simulations for testing communication effectiveness within business unit or installation per time period X – Level of standardization of communication about the job(s) within an installation or a business unit	– Percentage of employees with personal communication device of company (e.g., GSM) – Intensity and level of communication between shifts – Intensity and level of communication within shifts	– Level of standardization of communication terminology within company – Degree to which automated alarm systems are present – Degree to which alarm management is present in the company – Number of simulations for testing communication effectiveness within the company per time period X – Level of communication between company management and business unit management – Level of communication between different company departments (e.g., HR, production and SHEQ) – Degree to which intranet is used within the company – Degree to which layout facilitates communication

5. Learning facilities/ possibilities, training and education	– Degree to which legal requirements and industrial codes of good practice are achieved as regards training (considering task description of the function) – Degree to which task description synchronizes with training demands – Training/education costs per shift member – Number of double competences within shift – Number of multiple (triple, *etc.*) competences within shift – Time available per time period for training or education of function X – Number of training sessions fixed for function X	– Number of persons with competence X (concrete, for every competence) within shift of type Y (e.g., 4-shift system) – Average time period during which one shift member holds the same competences – Percentage of employees requiring crucial skills (X, Y, *etc.*) – Degree to which employees observe each other and coach each other in performing practices and carrying out tasks	– Percentage of managers trained in management skills – Training/education costs per manager – Number of double competences within installation or business unit – Number of multiple (triple, etc.) competences within installation or business unit – Time period after which competence or training session X is re-evaluated – Level of objectiveness and level of specificity of evaluation procedures for training sessions and education programs – Degree to which training sessions and education programs are repeated (e.g., for the case of non-standard circumstances) – Degree to which competence driven education is stimulated by management – Degree to which standards and codes of good practice are elaborated – Degree to which instructors and teachers are professionals – Number of simulation exercises per number of accident scenarios within a business unit or installation – Percentage of employees trained in crucial skills (X, Y, *etc.*) – Efficiency and completeness with which training schemes, competence profiles, and education programs are kept up-to-date

Table 7.7 Non-exhaustive list of concrete manning-level indicators and their objectives (illustrative example).

Concrete manning-level indicators (measured manning levels)	Concrete manning-level objectives (ideal manning levels)
Training of function A	Training X, Y, and Z for an installation of complexity Q
Criteria for changing/adapting lists of tasks	Criteria X, Y and Z for lists of tasks of type T
To be able to safely carry out action X	Exercise per period R for employee with training S
Monitoring level of satisfaction within shift	Carry out a satisfaction survey per period P
Number of periodic meetings within an organization to discuss manning-level issues	Number of meetings per period Y for business unit of size G
Training time yearly available per shift member	Number of days
Number of persons within a shift with competence X	Number of persons

objectives for every concrete manning-level indicator. Table 7.7 offers a simple example of such an exercise.

Suppose that a concrete manning-level objective is to hold Z meetings per period Y. A concrete figure (i.e. the result of measuring the concrete manning-level indicator) which deviates to a certain extent from this objective "Z" might indicate a sub-optimal situation exists.

If there is a change in manning-level demands, the (concrete and management) manning-level objectives will be changed. Based on these alterations, manning-level resources changes follow and the results of (concrete and management) manning-level indicators will be changed as well.

An action plan (see Figure 7.7) is suggested to put forward a systematic way to employ the different concepts and techniques discussed in the previous sections in this chapter. The action plan can be used on the company-level or on the level of a part of the company. The successive steps of the action plan to be followed by company manning-level management explain how to introduce, to elaborate, and to implement the operational manning-level optimization methodology.

7.7 Summary and Conclusions

Operational staffing-level optimization is a very complex matter including various disciplines and several departments within a chemical organization. Moreover, the

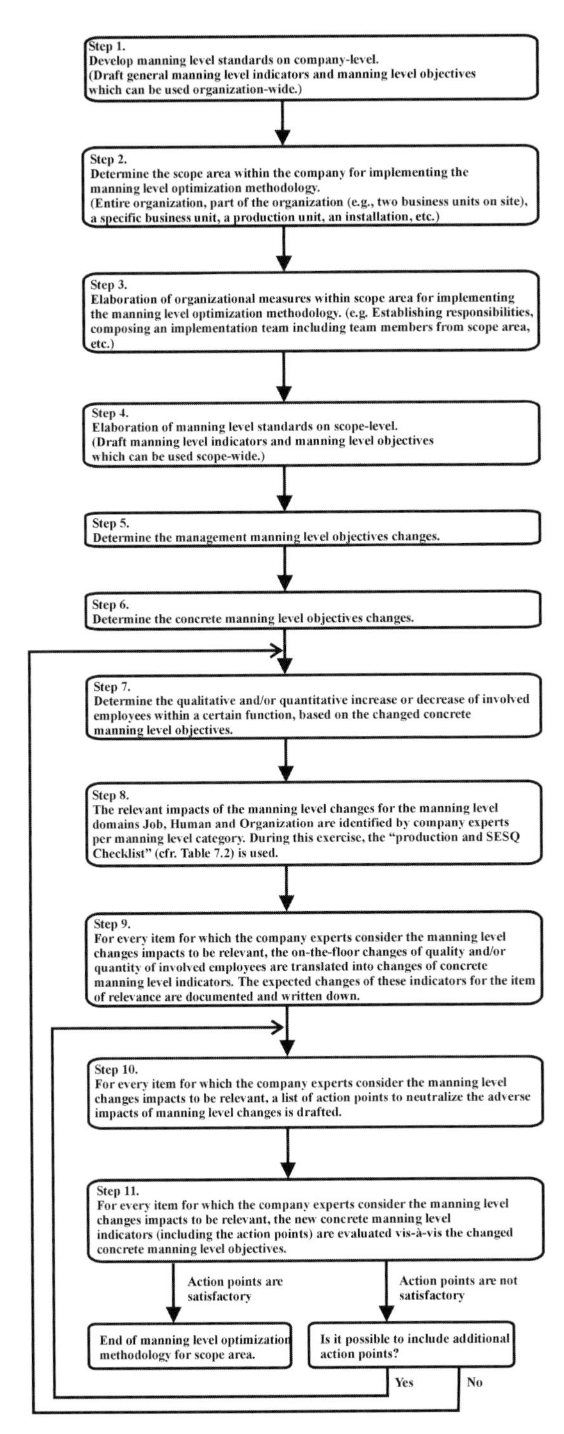

Figure 7.7 Staffing-level optimization action plan for a simple company. (The action plan can also be used by a multi-plant council, be it somewhat adapted.)

variety in organizational cultures and organizational features (size, activity, *etc.*) is huge. It is obvious that developing a generic management of change approach for assessing, evaluating and improving operational staffing levels in chemical plants is not at all an easy task. In this chapter, such an approach is elaborated and clarified.

As a part of the proposed approach, an instrument (a questionnaire) called IESLA that can be used by companies to evaluate the quality of their existing staffing levels and small staffing-level changes is suggested. A risk-based method (MCSL) for assessing the impact of large manning-level changes is proposed in the approach as well. The MCSL methodology can be employed by single-plant as well as by multi-plant management to prove that the impacts of the staffing-level changes are objectively detected and that appropriate measures are adopted to (continue to) meet all the production and/or SESQ requirements.

A stepwise plan of implementation (called the roadmap for staffing-level assessment) of the approach is set up. The suggested roadmap and its constituent instruments, checklists and methodologies allows its user to systematically and analytically assess and if required improve the operational staffing levels of a multi-plant industrial area, a chemical facility, an installation or business unit, whenever and wherever desired or required.

Furthermore, the well-known plan-do-check-act loop of continuous improvement is integrated into the methodology to help company management gaining insight into the optimization process. An action plan is provided for enabling easy implementation of the continuous optimization exercise.

If the approach is appropriately implemented, one can easily assume that the operational staffing levels are professionally and efficiently managed.

8
Multi-Plant Site-Integrated Safety and Security Governance

8.1
Introduction

Gaining multi-plant site-integrated safety and security is actually a twofold task: first a multi-plant site-integrated safety and security culture has to be established, and second it has to be maintained, continuously optimized and sustained.

Three capabilities for setting up a multi-plant-safety and -security culture were elaborated in the previous chapters, namely multi-plant-management procedures (Chapter 4), multi-plant management frameworks (Chapter 5) and multi-plant technology and tools (Chapter 6). These main building blocks have to be integrated to achieve an effective and efficient implementation in a multi-plant setting of chemical companies. The multi-plant council is then responsible for efficiently elaborating the entanglement of the building blocks thereby involving many different stakeholders with a variety of interests. Integration guidance should be independent of the size of the multi-plant area (e.g., the number of plants, the surface area or the amount of hazardous materials situated within the area). It should be applicable to every multi-plant situation in terms of legal requirements (taking into consideration differences between national legislations existing worldwide) and regardless of differences in single-plant safety and security cultures. Integration guidelines thus need to be abstract but nevertheless understandable to a variety of different people, situations, regulations, *etc.*

The methods and principles developed in previous chapters are evoked to support the discussion on the integration of the building block guideline documents.

8.2
From Individual Plant Safety and Security Know-How to Multi-Plant Safety and Security Knowledge

Plant safety and security know-how, the capacity to keep the chemical plant as safe and as secure as possible, remains the key to individual plant safety and security. However, due to the ever-increasing amount of hazardous chemicals processed at

Multi-Plant Safety and Security Management in the Chemical and Process Industries. G.L.L. Reniers

ISBN: 978-3-527-32551-1

the cluster level and the growing public perception that the industry should be considered safe and secured, the particular safety and security needs required to satisfy every plant's individual safety and security situation are ever more dependent on the neighboring plants as well. Therefore, individual plant S&S know-how should be aggregated and upgraded to multi-plant S&S knowledge. The individual plants have to develop, collect, filter, save, distribute and apply S&S knowledge processes at a multi-plant level. To attain this goal, harnessing, enhancing and supporting the common-knowledge processes of a group of chemical corporations has to be done intentionally and effectively by a supra-plant organization, in this book referred to as the multi-plant council. This so-called multi-plant council should thus be responsible for managing collective safety and security knowledge.

To efficiently manage multi-plant safety and security knowledge, it is of prime importance that individual company top management recognizes that mutual S&S knowledge generation, sharing and application contribute to the long-term success of corporations, individually, as well as clustered. Moreover, the way multi-plant safety and security management will need to operate is different from the way individual plant safety, health, environment & security departments operate at present. Promoting single-plant safety and single-plant security takes place on both the visible work floor where safety and security issues are implemented by the organization and on the (less transparent) intangible workplace where company safety and security proposals and ideas are being suggested. Since an immediate reciprocal link exists between these two types of workplaces in a company, the efficiency of implementing the continuous safety and security improvement cycle by plant safety and security management is relatively easy. The knowledge process can be thought of as (individual) plant learning.

Enhancing multi-plant safety and security is performed mainly on a (hidden) intangible supra-organizational level where multi-plant safety and/or security ideas are conceptualized. Afterwards, notions and procedures have to be concretized and implemented by parts of the multi-plant-organization, that is, organizational structures that adapt them to individual plant level.

Since multi-plant safety and security is highly dependent on the know-how and the knowledge of the plants forming the multi-plant safety and security area, single learning concerning plant safety and security needs to evolve into aggregated learning concerning multi-plant safety and security. Therefore, an intangible platform where multi-organizational S&S knowledge is created, shared and used, has to be established. The existing intangible plant S&S workplaces, which can be thought of and organized as plant departmental think tanks (see Chapter 4), can be combined and by cooperating they can serve as an input for an intangible multi-plant safety and security platform. The different multi-plant areas' long-term safety and security successes and leading-edge positions will depend on their capacity for joint learning. The concept for such aggregated learning is presented in Figure 8.1.

At the level of plant single-loop learning, plant safety and security managers try to optimize the performance of the same safety and security tasks by continually evaluating their outputs. The double-loop optimization comes into practice when safety and security managers reflect on the reasons why these tasks are being

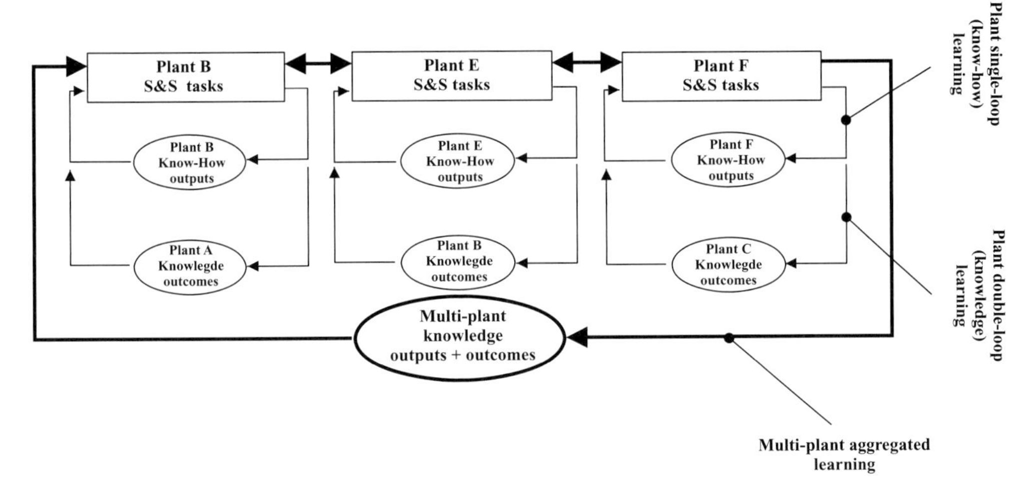

Figure 8.1 Multi-plant aggregated learning (e.g., for 3 plants *B*, *E* and *F*).

performed and critically question the aims and objectives. Multi-plant aggregated learning is characterized by gathering knowledge as a group of organizations by questioning the issues on individual plant level that can affect the entire group or parts of the group. It is obvious that aggregated learning can only be achieved by intensive cooperation based on solid guidelines, procedures and frameworks. Chapters 4, 5 and 6 elaborate these concepts and aim specifically at improving multi-plant safety and security.

Aggregated learning depends on the efficiency of intangible S&S workplaces. To establish these intangible multi-organizational S&S platforms, cutting-edge knowledge management (KM) technologies may be used. Inspired by KM, processes resulting in the coordination of cross-company cooperation should be elaborated. In order to work out these coordination processes,[1] it is necessary to define the individual roles, processes and technologies that can be combined in numerous ways to create unique KM solutions for a specific multi-plant area of chemical companies. This can be accomplished by setting up a knowledge network, for example, the network depicted in Figure 8.2.

By using similar networks, knowledge sharing can be developed and promoted within the multi-plant area. This, in turn, leads to collaborative technologies, people networks, and social exchanges for knowledge sharing. By safety and security data sharing, the level of safety and security innovation can be raised per plant, existing safety and security knowledge can be retained and leveraged per plant, safety and security knowledge transfer can be accelerated within the multi-plant area, best safety and security practices within the area can be identified and effectively deployed, safety- and security-related problem solving can be speeded up,

1) Coordination processes can be viewed as patterns of simpler dependencies and activities that describe a mechanism for managing the relationship implied by the dependency.

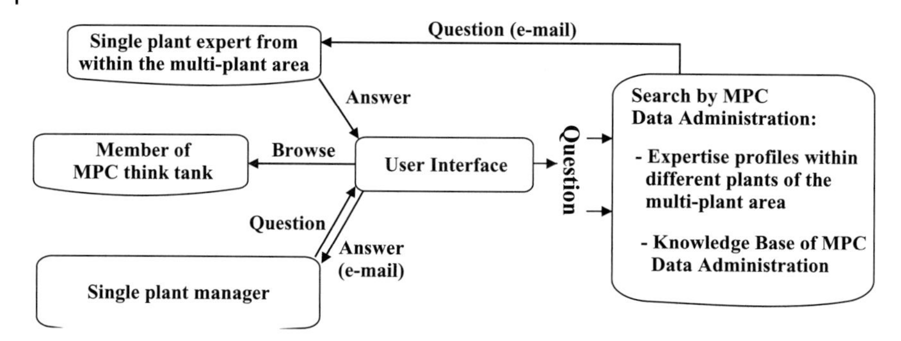

Figure 8.2 Knowledge network system architecture (Source: based on Rao, 2005).

new safety and security expertise can be integrated and exploited and–of course–aggregated learning about safety and security issues is accelerated. Coordination processes can thus be organized by implementing collaboration software, making the multi-plant area safety and security network a shared space where people can work together and by doing so, achieve step-by-step goals (e.g., multi-plant safety and security activities) resulting in continuous safety and security improvement cycles at multi-plant level as well as at single-plant level. Coordination processes represent an important part of multi-plant safety and security-capability integration, as will become clearer in the next section.

To efficiently manage all the concepts proposed in this book, a highly complex management design, which needs to be tailor-made for every multi-plant area, has to be drafted by the multi-plant council and implemented by all stakeholders involved. The next section makes powerful suggestions for this kind of design elaboration.

8.3
Towards a Design Code of Good Practice for Integrating Multi-Plant-Safety and -Security Building Blocks

As the size and the complexity of chemical clusters grow, the identification and correct management of interconnection dependencies among the building blocks of the safety and security culture (described in Chapters 4, 5 and 6) and the change management of staffing levels in chemical plants (described in Chapter 7) become increasingly and correspondingly important. The focus of the design effort lies in integrating the different building-block components by identifying and properly managing their interdependencies and mismatches.

The problem of connecting the building blocks is related to concepts such as resource flows, resource sharing and timing dependencies. Malone *et al.* (2003) indicate that the design of associated coordination protocols involves a set of mechanisms such as shared events, invocation mechanisms and communication protocols. Therefore, the multi-plant-safety and -security culture elaboration can be captured in a design space that assists designers belonging to the MPC to construct a coordination process that manages a given dependency type, simply by selecting the value of a number of design dimensions. To design and implement a solid safety and security culture within a multi-plant environment of chemical facilities (the multi-plant area), all activities related to multi-plant SMS/SMP, procedures for assessing staffing levels in the area, procedures for evaluating critical safety and/or security staffing arrangements, multi-plant management frameworks, multi-plant technology, *etc.*, need to be integrated. To streamline this very complicated management problem, activities need to interconnect with other activities, either because they use resources produced by other activities, or because they share resources with other activities. Moreover, the interconnection needs to be performed on three different levels: intra-company, cross-company and supra-company.

The development of a *design code of good practice* aims to reduce the specification and implementation of M-PSSC building blocks' interdependencies to a routine design problem, capable of being assisted, or even automated, by computer tools in the long run.

In order to propose these design guidelines that deal with the integration of multi-plant safety and security building blocks, it is first imperative to identify the people, activities and resources involved in establishing and managing a M-PSSC. Figure 8.3 offers an example of a non-exhaustive listing.

The next step is to find a method that represents all these factors and their associations with each other in a single diagram. To achieve this task, a diagram can be drafted for both safety and security, representing how to manage the multi-plant-management procedures continuous improvement design cycle (see Figure 8.4a,b).

In order to establish a general guideline to address multi-plant building block interconnection problems to be used by a variety of companies, it is necessary to make the important choice of the generic dependency types. Three dependency types can be identified (Malone *et al.*, 2003):

a) *Flow dependencies* encode relationships between producers and consumers of resources. Coordination protocols for managing such dependencies decompose into protocols that ensure accessibility of the resource by the consumers, usability of the resource and also synchronization between consumers and producers.

b) *Sharing dependencies* represent relationships among consumers who use the same resource or producers who produce for the same consumers. Coordination protocols for such dependencies divide a resource among different users or enforce mutual exclusion.

c) *Timing dependencies* express constraints on the relative flow of control among a set of activities. Such dependencies are used in the decomposition of cooperation protocols for flow and sharing dependencies.

To address the drafting of a dependency diagram, the resource flow graphs should be changed into dependency diagrams in which the focus shifts to just those flows that seem to be critical dependencies. Note that resource flow and flow dependency

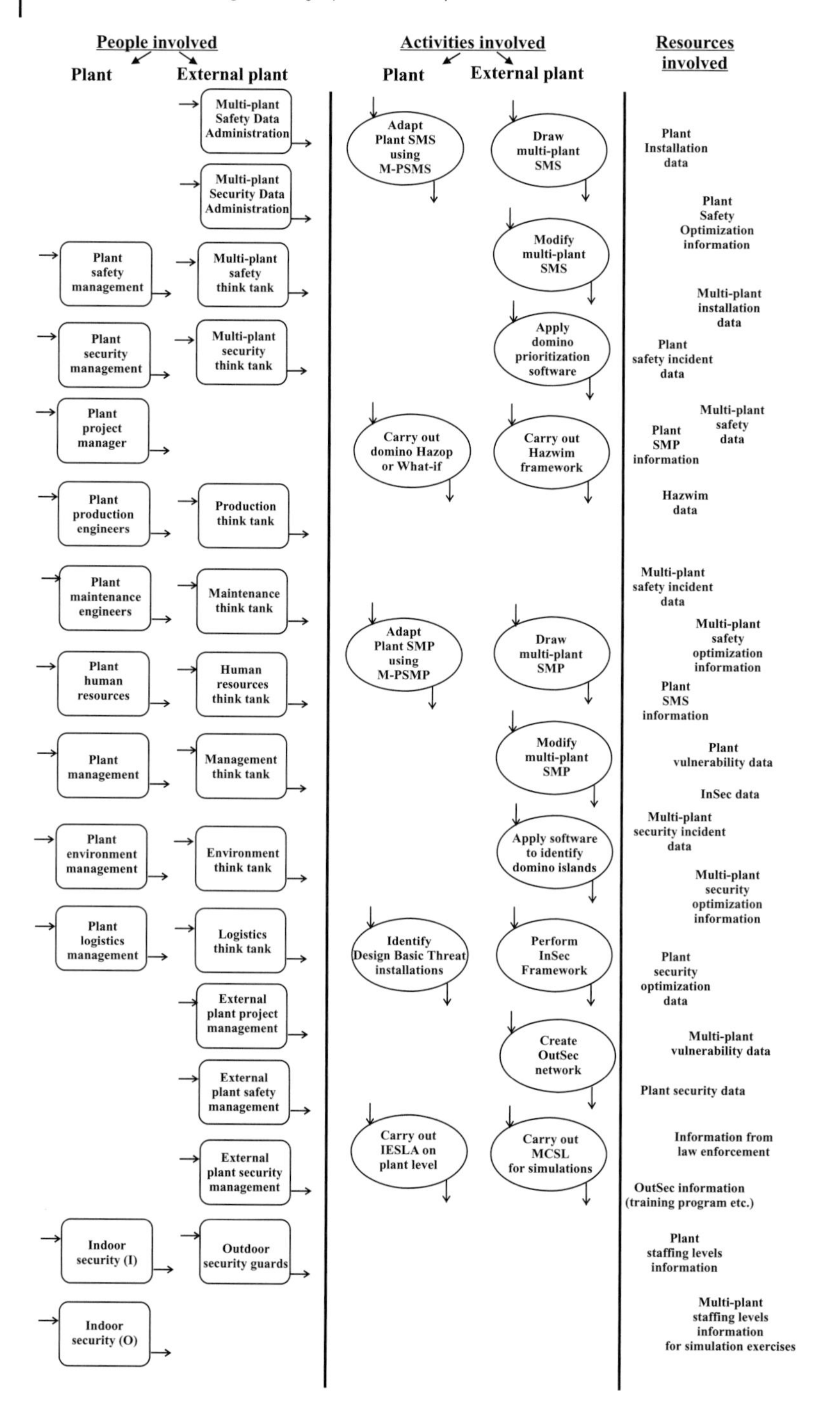

Figure 8.3 Non-exhaustive identification list of people, activities and resources for setting up a M-PSSC.

are isomorphs. This suggests that a dependency diagram could be produced simply by replacing each resource flow with a flow dependency. However, several issues are to be considered, for example, flows into a common activity may indicate a timing dependency; flows out of a common activity may indicate a sharing dependency, *etc.* Moreover, unlike a resource flow graph typically representing all the flows involving resources relevant to the activities shown, the dependency diagram represents only those dependencies that are important to the process; these critical dependencies correspond to a sub-set of the flows in the corresponding resource flow graph.

Note that in this sense a dependency diagram represents a further abstraction and hence simplification of a resource flow graph, the "importance" of a dependency being relative to the point of view and judgment of the process observer. To draft this kind of diagram, all the flow, sharing and timing dependencies that are implicit in the resource flow graph are drawn. Next, those dependencies that are unimportant to the observer are removed. A dependency might be considered unimportant for several reasons, for example, its effects on the process outcomes of interest are insignificant or it has a significant effect but is easily managed. Such dependency diagrams are illustrated in Figure 8.5a–c.

Figure 8.6 illustrates the merging of these diagrams, that is Figure 8.5a–c, into one simplified overall design process dependency diagram for the establishment of a multi-plant safety and security culture in an industrial area. The dependency diagram shown in Figure 8.6 can be considered as a multi-plant culture design guideline emphasizing the inputs to the design process and the fit among them. To manage and cultivate the multi-plant culture, the multi-plant council should focus on the critical inputs in this design activity.

This dependency diagram illustrates the power of abstraction in the multi-plant culture design approach by logically structuring the essentials of elaborating a M-PSSC.

8.4 Planning for Safety and Security Sustainability

As Beloff *et al.* (2005) indicate, despite the chemical industry's efforts to improve environmental and social performance and gain the public's trust, the industry still faces huge hurdles in terms of public acceptance and confidence. Fundamental topics such as historical performance, the failure of risk assessments, the intrinsic nature of hazardous chemicals, and the industry's vulnerability to terrorist threats help to explain this mistrust. In this book, a way ahead for sustaining safety and security in chemical clusters is given. This work thus constitutes important academic research to pro-actively approach the extremely important matter of site-integrated safety and security management in clustered chemical companies. Streamlining cross-company safety and security collaboration further introduces new ideas in safety- and security-management practices in chemical industrial areas and enhances overall sustainability.

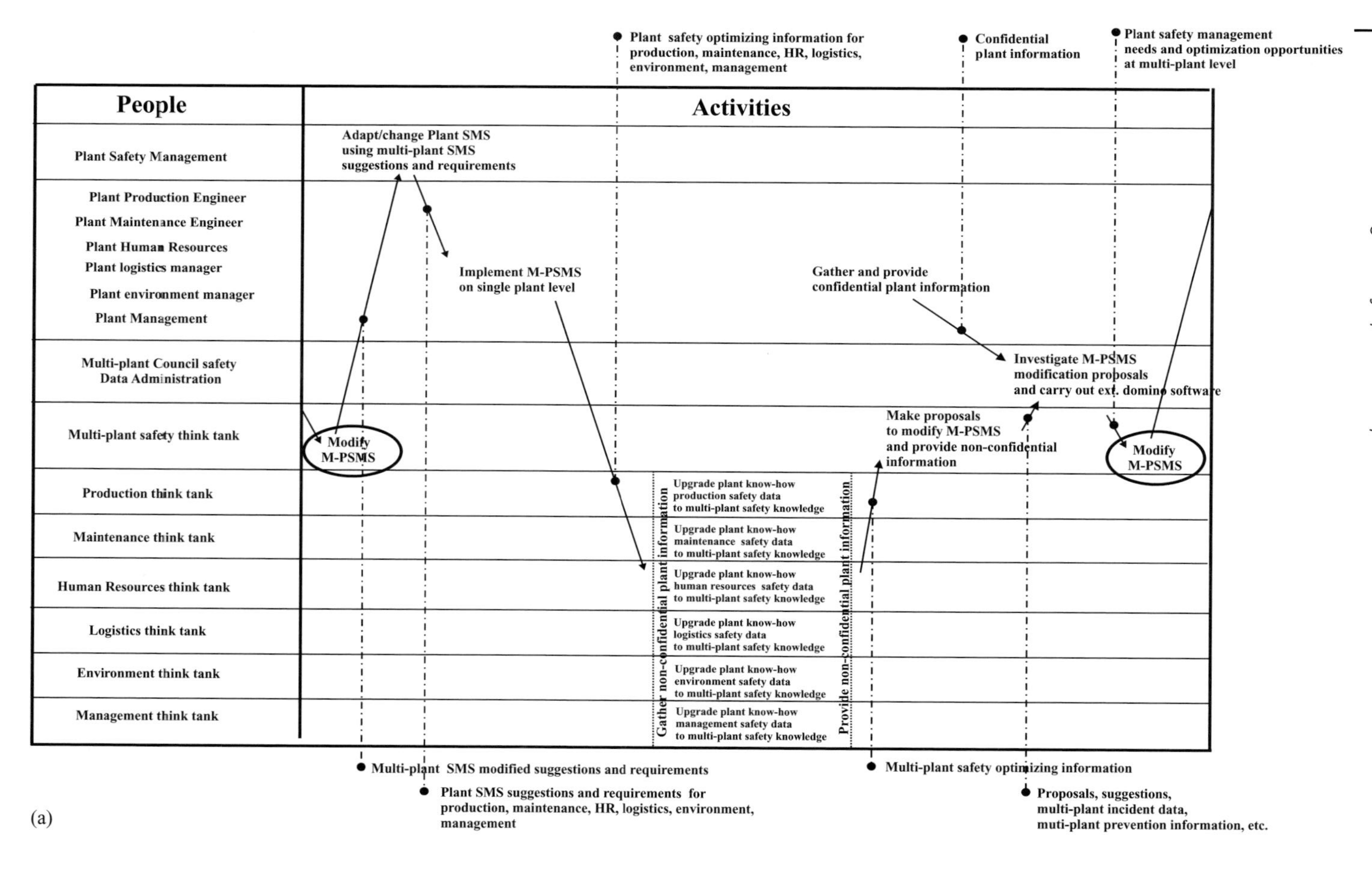
Plant safety optimizing information for
production, maintenance, HR, logistics,
environment, management
Confidential
plant information
Plant safety management
needs and optimization opportunities
at multi-plant level
People
Activities
Plant Safety Management
Plant Production Engineer
Plant Maintenance Engineer
Plant Human Resources
Plant logistics manager
Plant environment manager
Plant Management
Multi-plant Council safety
Data Administration
Multi-plant safety think tank
Production think tank
Maintenance think tank
Human Resources think tank
Logistics think tank
Environment think tank
Management think tank
Adapt/change Plant SMS
using multi-plant SMS
suggestions and requirements
Implement M-PSMS
on single plant level
Gather and provide
confidential plant information
Investigate M-PSMS
modification proposals
and carry out ext. domino software
Make proposals
to modify M-PSMS
and provide non-confidential
information
Modify
M-PSMS
Modify
M-PSMS
Gather non-confidential plant information
Upgrade plant know-how
production safety data
to multi-plant safety knowledge
Upgrade plant know-how
maintenance safety data
to multi-plant safety knowledge
Upgrade plant know-how
human resources safety data
to multi-plant safety knowledge
Upgrade plant know-how
logistics safety data
to multi-plant safety knowledge
Upgrade plant know-how
environment safety data
to multi-plant safety knowledge
Upgrade plant know-how
management safety data
to multi-plant safety knowledge
Provide non-confidential plant information
Multi-plant SMS modified suggestions and requirements
Plant SMS suggestions and requirements for
production, maintenance, HR, logistics, environment,
management
Multi-plant safety optimizing information
Proposals, suggestions,
multi-plant incident data,
muti-plant prevention information, etc.
(a)

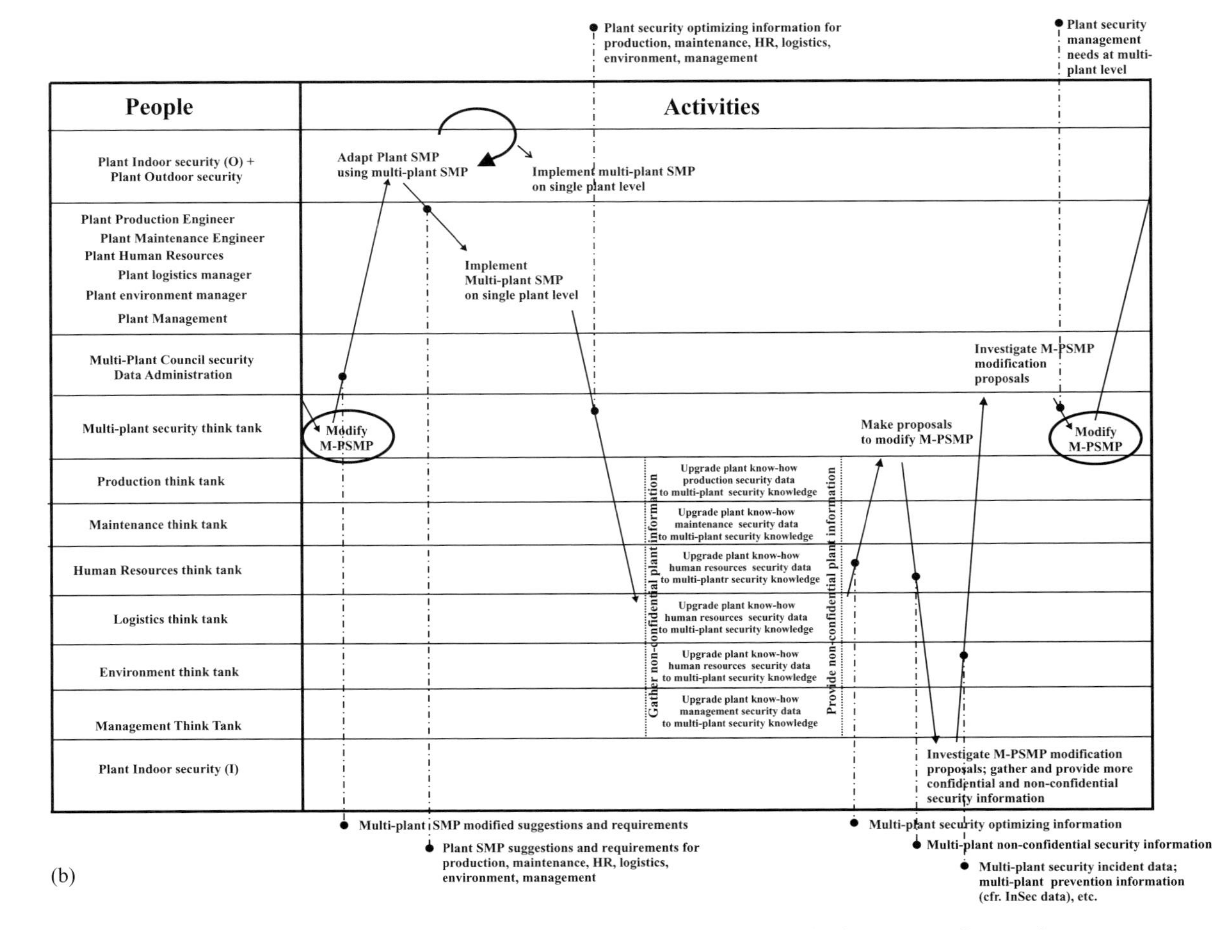

Figure 8.4 (a) Resource flow graph 1: M-PSMS continuous improvement design cycle; (b) Resource flow graph 2: M-PSMP continuous improvement design cycle.

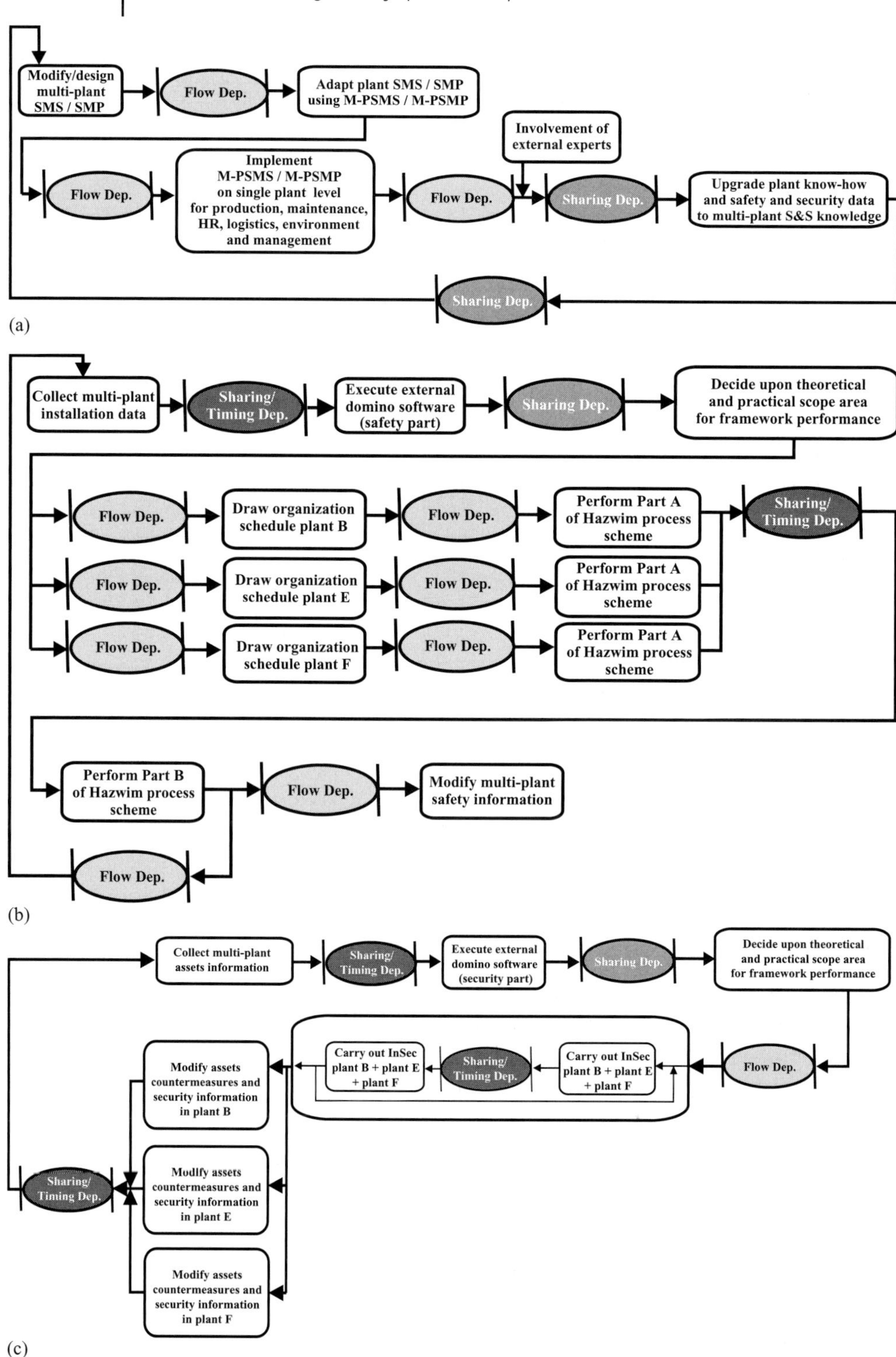

Figure 8.5 (a) A dependency diagram for the design cycle; (b) a dependency diagram for the multi-plant domino safety prevention improvement cycle; (c) a dependency diagram for the multi-plant domino security prevention improvement cycle.

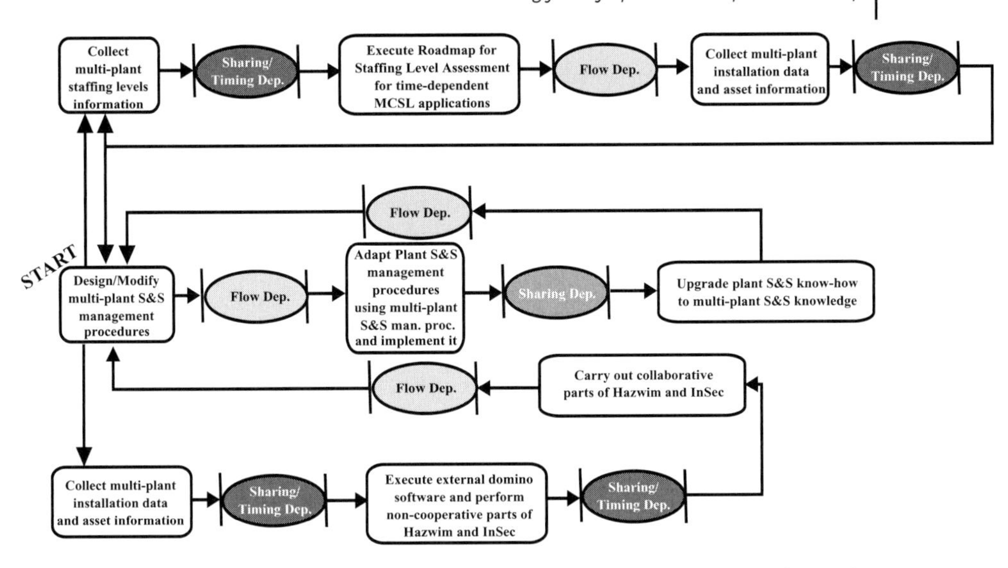

Figure 8.6 Simplified dependency diagram of designing a multi-plant culture in the chemical industry.

A sound safety and security culture transforms existing safety and security theories and practices to those that promote sustainability. Sustainable safety and security incorporates the development and implementation of cross-company techniques and the promotion of safety and security to all personnel, that is from personnel with the lowest level of responsibility within chemical plants to personnel having a holistic view of the chemical multi-plant area. It also includes taking decisions where there is a risk of serious or irreversible harm, such as in the case of external domino events, even in the absence of full scientific certainty. To fully implement sustainable safety and security solutions in chemical multi-plant areas, multi-plant safety and security management should use the following basic principles:

Principles for sustainable safety and security:

a) Implement the plan-do-check-act loop of continuous safety and security improvement for every function and at all levels within the chemical multi-plant area;

b) Strive to collaborate concerning safety and security theories and practices as much as possible/feasible;

c) Use the precautionary principle[2] to manage high-consequence, low-probability risks (such as external domino risks).

2) The precautionary principle can be described using the 1998 Wingspread Statement: "When an activity raises threats of harm to human health or the environment, precautionary measures should be taken even if some cause-and-effect relationships are not fully established scientifically."

Crucial for sustainable development in the chemical industry is the need to close the safety and security collaboration gap that currently exists in the industry. By implementing the suggestions, proposals, guidelines and recommendations made in this book, chemical multi-plant areas metabolize all kinds of accident complexities through a dynamic process of distributed learning that mobilizes the intelligence of their constituting separate plants that communicate, interact, and exchange products and/or information with each other. Since none of the individual plants has sufficient technical, regulatory or negotiating power to force the others to behave in accordance with a common action framework, each company learns in the course of action how to explore its own segment of accident complexity, accumulating experience itself and at the same time benefiting from the experience of others. The result is a circuit of self-organization that, through repeated attempts, leads to the development of a safety and security site-integrated multi-plant area with an efficient internal organization that can respond flexibly to the demands of multi-plant members and public authorities or to the demands of the community at large.

8.5
Summary and Conclusions

Few will disagree that the Responsible Care® program is one of the most renowned voluntary social and environmental initiatives launched by an industrial sector, in this case the chemical industry. However, Responsible Care®, while considered a component of sustainability (Beloff *et al.*, 2005), does not currently offer the cross-plant platform required to embrace multi-plant safety and security sustainability. Therefore, guidance on how to gain multi-plant site-integrated S&S, and how to sustain it, is made available in this chapter.

A simplified dependency diagram has been proposed to manage and control the M-PSSC building blocks. The diagram explains the integration of the components and the mechanism of the M-PSSC and explores an integrated solution for multi-plant building-block entanglement. It helps to automate multi-plant safety and security design and management practices and enhances the awareness of intra- and inter-company safety and security aspects for optimizing single-plant and multi-plant safety.

Associating each building block with coordination processes for managing it would reduce the process of improving the design of a M-PSSC to a routine. Prior to drafting a framework that encompasses and relates synchronization, communication and resource-allocation considerations, the multi-plant council has to introduce a technology for coordination processes (which is multi-plant specific). Hence, this leads to a two-step elaboration of the M-PSSC. First the MPC is encouraged to draft multi-plant-specific coordination processes using a software/service KM-based cluster-collaboration platform. Secondly, the MPC could develop the safety and security culture step by step according to the proposed dependency diagrams.

Reflecting on the implementation of plant and multi-plant safety and security, knowledge must be integral to the way in which safety managers and security managers come to make sense of continuously optimizing safety and security in a chemical industrial area. It will increasingly become a routine competency for making effective and (if necessary) rapid company and cross-company safety- and security-related decisions.

9
Game-Theory: A Mathematical Technique to Convince Company Top Management to Invest in Multi-Plant Safety and Security

9.1
Introduction

Persuading plant management to invest in external domino-effect prevention and starting up safety and security collaboration with neighboring chemical companies, is essential to develop a successful multi-plant initiative. Establishing an effective multi-plant council can only truly be achieved if top management from various chemical plants fully engage in multi-plant safety and security on a strategic level. However, no clear and straightforward economic incentives exist to urge companies within chemical clusters to jointly develop multi-plant external domino-risk management. To this end, using a broad-based game-theoretical perspective, this chapter is concerned with how to make the financial benefits of external domino-effects-prevention measures obvious to top-management of two adjacent chemical plants.

Since 2001, game theory has been used as a promising scientific technique to deal with security issues (Liu *et al.*, 2008; Golany *et al.*, 2009; Hausken and Levitin, 2009). Although game-theoretic modeling in combination with reliability theory has already been employed in scientific research to gain insights into the nature of optimal defensive investments that yield the best trade-off between investment costs and security of critical infrastructures (Bier *et al.*, 2005; Azaiez and Bier, 2007; Hausken, 2008), to date, no attention has been paid to the multi-plant safety- and security-related prevention decisions (concerning external domino effects) made by plant management of neighboring chemical companies. Nonetheless, these decisions have an important impact on whether or not an external domino effect might take place and/or what consequences can be expected.

Risks as regards external domino effects between two chemical plants are risks whose consequences depend on a company's own risk-management strategy and on that of the adjacent company. To develop an external domino-effect investment model in the case of two neighboring companies, the way in which a single chemical plant manages its external domino-effect risks where there is a likelihood that even if it has decided to invest in adequate measures, it might be harmed due to its neighbor not investing, needs to be thoroughly analyzed.

Multi-Plant Safety and Security Management in the Chemical and Process Industries. G.L.L. Reniers

ISBN: 978-3-527-32551-1

The research described in this chapter is thus aimed at predicting the external domino-effect prevention outcome of a situation where both companies make independent decisions on whether to invest in such prevention or not, but are at the same time aware of the strategic external domino-effect decisions (to "invest" or to "not invest") made by the other company. In real industrial settings it is obvious that chemical clusters are often composed of more than two companies. However, domino-effects prevention first needs to be focused on two companies, since these two plants actually might trigger the escalation effect, eventually involving more plants. Hence, if the triggering plants are able to contain the potential domino effect, no knock-on effect can affect the other plants within the cluster. Therefore, strategic external domino-prevention choices (which can be safety oriented as well as security oriented) of two companies situated next to each other are first investigated. Afterwards, the possibility of changing the strategic choice of a small number of companies of a cluster (composed of at least three companies) and to tip all the rest of the companies within the cluster to change their strategies from a socially non-optimal situation to a socially optimal situation is investigated.

Game theory is the theory of independent and interdependent decision making. Games of strategy are games involving two or more players, not including nature, each of whom has partial control over the outcomes. The external domino-effect game can be classified as a two-person mixed-motive[1] game of strategy (Kelly, 2004; Barron, 2008). Even the simplest mixed-motive games, represented by two-by-two matrices, have many strategically distinct types. There are twelve distinct symmetrical two-by-two mixed-motive games, of which eight have single Nash equilibrium points[2] and four do not (Rapoport and Guyer, 1966). In the external domino-effect case, the game can be classified as a mixed-motive game without a single equilibrium point (but characterized with more than one equilibrium point). One archetype of such a game is a so-called "martyrdom game".[3] The most famous prototype of such game type is the prisoner's dilemma game. Such a game is characterized with a conflict between individual self-interest and collective self-interest. A martyrdom game has one Nash equilibrium point. Hence, the external domino-effect-prevention investment choices made by every individual chemical facility might lead to socio-economic optimal or sub-optimal situations. If the costs are sufficiently low (so that each company wants to invest in external domino-effects prevention, even if the neighboring company did not incur these costs) is a straightforward example of a socio-economic optimal situation. If external domino-effects prevention investments would appear to be very

1) In a mixed-motive game, the sum of the pay-offs differs from strategy to strategy.
2) The Nash equilibrium concept embodies two requirements (Vega-Redondo, 2003): (i) players' strategies must be a best response (i.e. should maximize the players' respective payoffs or should minimize their respective costs), given some well-defined beliefs about the strategies adopted by the other players, and (ii) the beliefs held by each player must be an accurate *ex ante* prediction of the strategies actually played by the other players.
3) For specific situations where the external domino effects game is characterized by a costs matrix indicating it is not a martyrdom game, this specific game can of course be solved using the appropriate game-theoretical solving method.

high to both companies relative to their potential benefits, then it might be efficient for no company to incur investment costs (inducing a socio-economic sub-optimal situation). It should be noted that in the context of this book, a socio-economic optimal situation represents a situation where society and economy are protected as well as possible at the optimal (minimized) cost against the devastating consequences of a major accident, *in casu* an external domino effect.

Due to the extremely low probabilities of an external domino effect occurring, many chemical plants are not inclined to invest in measures to prevent such accidents. If this is the case, companies believe that, whether their neighbor invests or does not invest in such measures, the companies' strategy to "not invest" is always better than "to invest". Hence, in current industrial practice, in the external domino-effect game played between two neighboring chemical plants, the Nash equilibrium seems to be for both companies to follow a strategy to "not invest" in external domino-effects prevention.

However, in the external domino-effects game the equilibrium representing both companies not investing is possibly socio-economically worse than another possible strategy where both players agree to invest, due to the conflict of individual self-interest and collective self-interest. The strategy where both companies minimize their collective costs is, however, unstable, since each player is tempted to deviate from it (due to the very low probabilities of an external domino effect occurring, in combination with high investment costs to avoid such an accident).

9.2 Qualitative Discussion on Multi-Plant-Safety and -Security Investments

If a company decides to invest in multi-plant safety and security, not only the company itself, but also its neighboring company is less vulnerable to an incidental domino effect or to a terrorist attack triggering an external domino effect. Expectations and perceptions about the neighbors' decisions will thus influence investments in multi-plant safety- and security-prevention measures. As a result of perception, the socio-economic outcome might be sub-optimal for both companies. As mentioned in the introductory section, this situation of decision making of two neighboring plants can be modeled as what is called a "game" and – by solving the game – give conditions for a win-win situation where both companies win by investing in multi-plant safety- and security-prevention measures. The financial background of the choice "to invest or not" is illustrated in Figure 9.1.

The expected losses *versus* performance curve depicted in Figure 9.1 measures the expected losses as a result of multi-plant safety and security precaution investment decisions *versus* the performance of these precautions (representing safety and/or security of the company). The curve should be upward sloping because the higher the expected (investment) losses (expected losses are seen here as possible costs due to following a certain investment strategy) are, the more or better safety and security precautions within the multi-plant area, and thus the

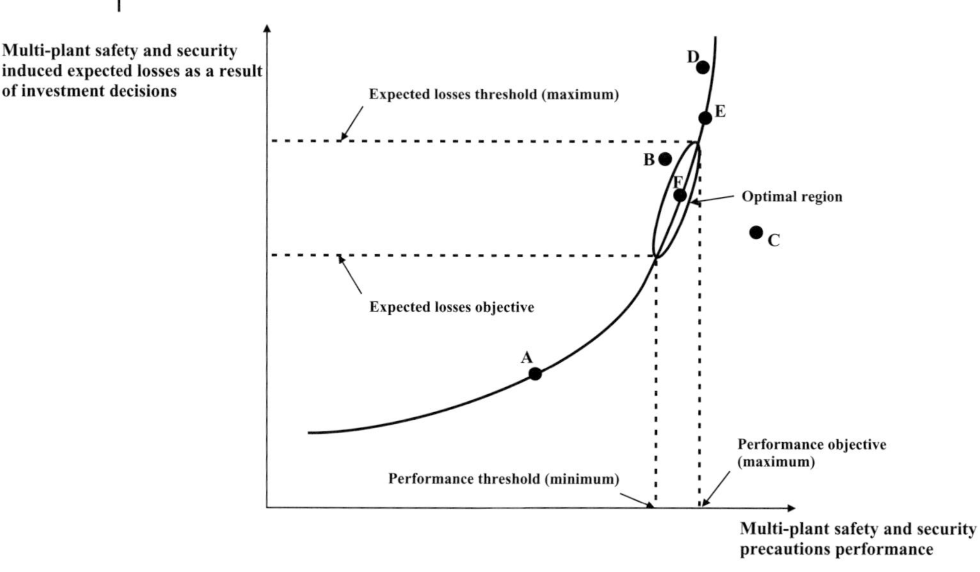

Figure 9.1 Multi-plant safety and security prevention measures planning in the case of two neighboring chemical companies (Source: based on Conrow, 2003).

higher the precautions' performance should be. The higher the multi-plant safety and security precautions' performance is desired, the higher the costs to reach this performance. Points on the curve indicate an efficient planning of precautions: no increase in performance can be obtained without a corresponding increase in expected losses, and no decrease in expected losses can be obtained without a decrease in performance. However, points on the expected losses *versus* performance curve do not necessarily represent optimal planning. Point A in Figure 9.1 illustrates the latter: although multi-plant safety and security precautions are planned very efficiently, their performance is not adequate (below the minimum performance threshold), thus planning is not optimal. Points lying above (or to the left of) the curve indicate feasible, but inefficient, planning. This is indicated by point B in Figure 9.1. Here, the planning is an inefficient combination of expected losses and performance, because multi-plant safety and security prevention could be developed at the same expected loss but with higher performance or at the same level of performance with lower expected losses (e.g., by cooperating with neighboring companies through using a game-theoretical approach described in this chapter). Points below (or to the right of) the curve (point C in Figure 9.1) indicate infeasible planning for a given set of input constraints (e.g., functional safety/security and/or safety/security limits, technology level limits, safety/security program limits, *etc.*).

Most companies are unwilling to invest in multi-plant safety and security precautions in present industrial settings because they assume that they are situated at point D (multi-plant S&S measures taken individually) or point E (multi-plant S&S measures taken by collaboration) of the expected losses *versus* performance curve. Hence, this would mean a maximum multi-plant safety- and security-prevention performance, but at very high expense in excess of the companies' expected losses thresholds. The optimal multi-plant S&S precautions planning should be at an acceptable expected loss with an acceptable performance (point F in Figure 9.1).

Managing multi-plant S&S and prevention investments involve different companies and thus different decision makers. Therefore, using the game-theoretical approach further explained could prove a means of meeting the objective of jointly reaching an acceptable performance at an acceptable expected loss for multi-plant preventive measures.

9.3 Two-Plant External Domino-Effects Investment Model

Following Heal and Kunreuther (2007), we classify the game as a game of partial protection with negative externalities, whereby we define "externalities" as possible effects that one company can have on another company. The existence of possible external domino effects between two companies gives rise to negative externalities. An accident caused by a lack of domino-effect prevention within one company can have catastrophic effects (i.e. "negative externalities") on its neighboring company. A Nash equilibrium exists for this kind of game-theoretic problem.

Furthermore, both companies have a discrete strategy, S_i (i = 1 or 2), that can take as values either I or NI, representing investing in external domino-effects prevention (I) and not investing in external domino-effects prevention (NI), respectively. Consider two adjacent chemical companies. Let the factor $P_{i,j}$ represent the probability that an accident will occur in plant j caused by an accident that took place in plant i (in other words, $P_{i,j}$ is the likelihood of an external domino effect from company i to company j). If $i = j$, then the factor expresses the probability for an internal domino effect in company i. Every company can decide either to invest in external domino-effect prevention or not. If company j does not invest, the pure investment cost of company i equals $c_i^{j\mathrm{NI}}$. If company j does invest, the pure investment cost of company i equals c_i^{jI} , with $c_i^{jI} < c_i^{j\mathrm{NI}}$ (due to possible benefits and efficiencies[4]). If an external domino accident takes place, the loss to company i equals L_i. For simplicity, external domino-effect-prevention measures are assumed to be completely effective. Hence, if external domino-effect-prevention investments are made in company i, no domino effect can originate from company i towards company i itself, as well as towards company j.

To investigate whether it is possible in the two-company case study to obtain a socio-economic optimal situation of both companies investing in external domino-effect prevention, the conditions for obtaining a Nash equilibrium point have to be established. Therefore, the costs matrix of the game has to be drawn.

4) Both companies can agree on the lowest joint investments for adequate prevention and protection against external domino effects.

Company 1 \ Company 2	*I*	*NI*
I	c_1^{2I} ; c_2^{1I}	$c_1^{2NI} + P_{2,1}L_1$; $P_{2,2}L_2$
NI	$P_{1,1}L_1$;$c_2^{1NI} + P_{1,2}L_2$	$P_{1,1}L_1(1-P_{2,1}) + P_{2,1}L_1(1-P_{1,1})$; $P_{2,2}L_2(1-P_{1,2}) + P_{1,2}L_2(1-P_{2,2})$

Figure 9.2 Costs matrix of companies 1 and 2 for the external domino-effects game.

Assume two companies called 1 and 2. If both companies 1 and 2 decide to invest in external domino-effects preventive measures, then their costs are just their investment costs, c_i^{jI} with $(i, j) = (1, 2)$ and $(2, 1)$. If company 1 invests in external domino-effects prevention measures, and company 2 does not invest, then company 1 incurs its investment cost c_1^{2NI} plus the expected loss of an accident which is initiated by an accident inside company 2 (i.e. $P_{2,1}L_1$). In such case, company 2 just has an expected loss from an accident initiated within its own company, that is, $P_{2,2}L_2$. An analogous cost allocation can be made for company 1 not investing and company 2 investing. If neither company 1 and company 2 invest, then company 1 has an expected loss from (i) an accident initiated within the own company (i.e. $P_{1,1}L_1$), conditioned on there being no accident from company 2 onto company 1 (i.e. times $1 - P_{2,1}$), plus (ii) the expected loss from an accident from company 2 onto company 1 (i.e. $P_{2,1}L_1$), conditioned on there being no accident initiated within the own company (i.e. times $1 - P_{1,1}$).[5] The conditions result from the fact that a chemical installation can only explode or be destroyed once and that the internal and external accidents do not originate at the same time. Direct annual external domino-effect prevention investment costs are assumed. The resulting costs matrix can be found in Figure 9.2.

For this costs matrix, the strategy {*NI*, *NI*} is a Nash equilibrium point under the conditions that:

$$\begin{cases} P_{1,1}L_1(1-P_{2,1})+P_{2,1}L_1(1-P_{1,1}) < c_1^{2NI}+P_{2,1}L_1 \\ P_{2,2}L_2(1-P_{1,2})+P_{1,2}L_2(1-P_{2,2}) < c_2^{1NI}+P_{1,2}L_2 \end{cases} \tag{9.1}$$

If these conditions are satisfied, companies 1 and 2 do not have an incentive to deviate from their strategy NI since, whatever strategy the other company has chosen, the Nash equilibrium is optimal (i.e. the costs are lowest). In Figure 9.2, the Nash equilibrium point is indicated in a gray rectangle.

5) Note that the expected loss of company 2 for this situation can be calculated in a similar way.

The strategy NI is for both chemical enterprises a dominant strategy[6] under the conditions that:

$$\begin{cases} P_{1,1}L_1 < c_1^{2I} \\ P_{1,1}L_1(1-P_{2,1}) + P_{2,1}L_1(1-P_{1,1}) < c_1^{2\mathrm{NI}} + P_{2,1}L_1 \\ P_{2,2}L_2 < c_2^{1I} \\ P_{2,2}L_2(1-P_{1,2}) + P_{1,2}L_2(1-P_{2,2}) < c_2^{1\mathrm{NI}} + P_{1,2}L_2 \end{cases} \tag{9.2}$$

Furthermore, the external domino-effects game is a martyrdom game under the conditions that:

$$\begin{cases} P_{1,1}L_1 < c_1^{2I} < c_1^{2\mathrm{NI}} + P_{2,1}L_1 \\ P_{1,1}L_1 < P_{1,1}L_1(1-P_{2,1}) + P_{2,1}L_1(1-P_{1,1}) \\ P_{2,2}L_2 < c_2^{1I} < c_2^{1\mathrm{NI}} + P_{1,2}L_2 \\ P_{2,2}L_2 < P_{2,2}L_2(1-P_{1,2}) + P_{1,2}L_1(1-P_{2,2}) \end{cases} \tag{9.3}$$

These conditions are satisfied if the probabilities of an external domino effect have such extremely low values that $c_i^{jI} > P_{i,i}L_i$. If the external domino-effects game is a martyrdom game, both companies have dominant strategies and one Nash equilibrium point. If one company deviates from the Nash equilibrium point, it suffers itself (i.e. it becomes a martyr) and benefits the adjacent company. If both neighboring companies deviate from the Nash equilibrium point, the cost is lower for both ("martyrdom equilibrium point"). In other words, the deviating company from a martyrdom equilibrium point moves to point D or E of Figure 9.1, and the other company remains situated in point B of Figure 9.1. If both neighboring companies deviate from the Nash equilibrium point, the pay-off is better for both (as they end up in the martyrdom equilibrium point). In this case, both companies are situated in the optimal region (e.g., point F) of Figure 9.1. Note that the adjacent companies may communicate with each other if they so choose in this type of game (Aumann, 1989). It makes no difference. The companies might agree to invest in external domino-effects prevention before the game, but they will still choose selfishly to not invest when faced with the actual decision, if acting rationally.

Four theoretically possible scenarios can be distinguished that will be further discussed. Table 9.1 summarizes these possible scenarios.

Further details on the scenarios are given below.

9.3.1 Scenario 1

Scenario 1 represents a game with two stable equilibria, (*I*, *I*) and (NI, NI). This is a mixed-motive game of strategies and there is no dominant strategy in the game. If both companies proceed to invest, the risk of external domino effects is reduced to zero. Under these conditions, the cluster is in an optimal socio-

6) A dominant strategy is always an optimal strategy for a player, independent of another player's strategy (Montet and Serra, 2003).

Table 9.1 Possible scenarios and their attributes.

Scenario	Conditions	Martyrdom's game?	Stable equilibrium	Socio-economic optimal equilibrium
1	$c_1^{2I} < P_{1,1}L_1$ and $c_2^{1I} < P_{2,2}L_2$	No	(I,I)(NI,NI)	(I,I)
2	$P_{1,1}L_1 < c_1^{2I} < P_{1,1}L_1(1-P_{2,1}) + P_{2,1}L_1(1-P_{1,1})$ $P_{2,2}L_2 < c_2^{1I} < P_{2,2}L_2(1-P_{1,2}) + P_{1,2}L_2(1-P_{2,2})$	Yes	(NI,NI)	(I,I)
3	$P_{1,1}L_1(1-P_{2,1}) + P_{2,1}L_1(1-P_{1,1}) < c_1^{2I} < c_1^{2I} + P_{2,1}L_1$ $P_{2,2}L_2(1-P_{1,2}) + P_{1,2}L_2(1-P_{2,2}) < c_2^{1I} < c_2^{1I} + P_{1,2}L_2$	Yes	(NI,NI)	(NI,NI)
4	Other possible conditions	No	(NI,NI)	(I,I)(NI,NI)

economic situation. In scenario 1, equilibrium (I, I) investments are lower for each player than equilibrium (NI, NI) costs. Since the model is characterized with imperfect but complete information and is based on rationality,[7)] both players will opt for a strategy to invest. The game is no longer a martyrdom game and the socio-economic optimum is reached. There is, therefore, no reason to use additional incentives to encourage companies to carry out joint investments in external domino-effect-prevention measures in this scenario.

9.3.2
Scenario 2

In Scenario 2, the conditions are given for a martyrdom game and not investing is the dominant strategy for both players. If none of the companies invest (strategy = (NI, NI)), company 1 incurs a cost $P_{1,1}L_1(1 - P_{2,1}) + P_{2,1}L_1(1 - P_{1,1})$ and company 2 incurs a cost $P_{2,2}L_2(1 - P_{1,2}) + P_{1,2}L_2(1 - P_{2,2})$ per year. This is a stable Nash equilibrium. If one player deviates from this equilibrium, he becomes a martyr of the game. Nonetheless, the best socio-economic solution would be (I, I) since in that case companies only incur costs c_1^{2I} and c_2^{1I} per year, respectively, being lower than in the Nash equilibrium. This socio-economic equilibrium is unstable as both companies intend to deviate from it, since in the situation where one company deviates and the other does not the incurred costs of the deviating company are further decreased (due to $P_{1,1}L_1 < c_1^{2I}$ and $P_{2,2}L_2 < c_2^{1I}$). In order to obtain the socio-economic situation, that is (I, I), incentives may be given. Possible incentives are establishing a multi-plant council dealing with cross-plant safety issues, and/or awarding subsidies or demanding taxes, and/or stimulating or discouraging investments with insurance-premium fluctuations.

7) In game theory, rational behavior refers to a consistency in decision making. Consistent players are seeking to maximize their pay-offs (Kelly, 2004).

As safety investments at multi-plant-level are characterized by game-theoretical features, a good way to enhance multi-plant collaboration and external safety investments is to adopt a supra-plant approach. In Chapter 4 the way to set up an institution at the cluster-level, the so-called multi-plant council, which would be responsible for a continuous follow-up of external safety and security improvements at the member companies, is explained. Due to its cross-plant trust inducing capability, the multi-plant council might play a stimulating role to reach the socio-economic optimum.

The socio-economic optimal situation (I, I) may also be stabilized through granting subsidies or by lowering the insurance premiums. The size of subsidy can be easily inferred from the cost matrix:

$$\begin{cases} c_1^{2I} - S_1 < P_{1,1}L_1 \\ c_2^{1I} - S_2 < P_{2,2}L_2 \end{cases} \tag{9.4}$$

where S_1 is the subsidy or lowered premium difference for company 1 and S_2 is the subsidy or lowered premium difference for company 2. The amount must cover at least the additional costs of safety investments compared with the expected loss of an accident. In this case, the companies will move from a socio-economic sub-optimal situation towards an optimal equilibrium and decide to invest. The game has now mixed strategies and two stable equilibria. Assuming rationality, the stable equilibrium (I, I) will result.

Tax imposition or raising the insurance premium can lead to a stable equilibrium (I, I) as well. The precise amount is also easily deduced from the cost matrix:

$$\begin{cases} c_1^{2I} < P_{1,1}L_1 + T_1 \\ c_2^{1I} < P_{2,2}L_2 + T_2 \end{cases} \tag{9.5}$$

where T_1 is the tax or raised premium difference imposed on company 1 and T_2 is the tax or raised premium difference imposed on company 2. The company will have to pay a tax or an increased premium if it does not meet the standards of a socio-economic optimum. When the investment costs minus the additional costs imposed on a company are lower than the expected loss of an accident, this company will opt for investing.

Obviously, the government may choose to employ a combination of subsidies and taxes to achieve the desired investment behavior. Similarly, insurance companies can also carry out a mixture of premium increase and reduction.

9.3.3 Scenario 3

In this situation the conditions are given for a martyrdom game. Both players will have the same dominant strategy, that is to not invest. In this way, a stable Nash equilibrium results: (NI, NI). The Nash equilibrium is also the most optimal socio-economic situation in scenario 3 as the costs in this situation will be minimal anyway. Accordingly, using incentives in such a scenario is unnecessary.

9.3.4
Scenario 4

Costs to collaborate (and to obtain strategy (*I*, *I*)) will be higher than non-cooperation costs in scenario 4 for at least one of the two companies. In this case (NI, NI) becomes the stable Nash equilibrium offering the lowest-cost option and the companies have no reason to deviate from it.

Depending on the case, the government or the insurance company may scrutinize if it is socio-economically responsible to give incentives to either (or both) of the companies whereby the costs of collaboration are reduced to a level where jointly investing is, taking the incentives into account, more advantageous to the companies.

9.3.5
Recommendations Based on the Scenarios

It is obvious that exchanging information between two adjacent companies regarding potential costs, accident probabilities and safety investments might be rewarding for both companies. This information would allow the plants to draw cost matrices in a simple way, whereby socio-economic optima and Nash equilibria can be determined. Building upon these findings, companies are then able to develop an optimal cross-plant safety and security policy, which benefits themselves as well as their surroundings.

Industrial application of the external domino-effects-investment model might thus lead to lower investment costs and at the same time bring about a truly safer chemical cluster.

9.4
Two-Plant Martyrdom Games

A formal solution of the martyrdom game can be found with the help of meta-game theory (Howard, 1966). Meta-game theory is the construction of any number of higher-level games based on the original game. A player is then assumed to choose from a collection of meta-strategies, each of which depends on what the other player chooses (Kelly, 2004).

In the external domino-effects game, company 1 has two pure strategies: "to invest in external domino-prevention measures" and "to not invest in external domino-prevention measures". For each of these, company 2 has four meta-strategies:

- To invest (*I*) in external domino prevention regardless of what company 1 does;
- To not invest (NI) in external domino prevention regardless of what company 1 does;
- To choose the same strategy as company 1 is expected to choose;
- To choose the opposite strategy to the one company 1 is expected to choose.

Company 1 \ Company 2	*I*, regardless of company 1	*NI*, regardless of company 1	Choose same strategy as company 1	Choose opposite strategy to company 1
I	$c_1^{2I};c_2^{1I}$	$c_1^{2NI}+P_{2,1}L_1;P_{2,2}L_2$	$c_1^{2I};c_2^{1I}$	$c_1^{2NI}+P_{2,1}L_1;P_{2,2}L_2$
NI	$P_{1,1}L_1;c_2^{1NI}+P_{1,2}L_2$	$P_{1,1}L_1(1-P_{2,1})+P_{2,1}L_1(1-P_{1,1});$ $P_{2,2}L_2(1-P_{1,2})+P_{1,2}L_2(1-P_{2,2})$	$P_{1,1}L_1(1-P_{2,1})+P_{2,1}L_1(1-P_{1,1});$ $P_{2,2}L_2(1-P_{1,2})+P_{1,2}L_2(1-P_{2,2})$	$P_{1,1}L_1;c_2^{1NI}+P_{1,2}L_2$

Figure 9.3 Costs matrix of companies 1 and 2 for the first-level multi-plant safety and security meta-game.

Figure 9.3 illustrates the costs matrix for this first-level meta-game.

In this two-by-four matrix representing the first-level multi-plant safety and security meta-game there is an equilibrium point at (row 2, column 2), since if company 1 chooses row 1 then company 2 should choose column 2 or column 4. In addition, if company 1 chooses row 2, then company 2 should choose column 2 or column 3. This corresponds to the same paradox point in the original game and thus a higher-level meta-game must be determined in order to solve the game.

Suppose company 1 selects a meta-strategy depending on which of the four meta-strategies company 2 chooses. Company 1 can then choose between 16 possible meta-strategies:

- To invest (*I*) in external domino prevention (EDP) no matter which of the four columns company 2 chooses;
- To invest (*I*) in EDP, unless company 2 chooses the fourth column;
- To invest (*I*) in EDP, unless company 2 chooses the third column;
- To invest (*I*) in EDP, unless company 2 chooses the second column;
- To invest (*I*) in EDP, unless company 2 chooses the first column;
- To invest (*I*) in EDP if company 2 chooses the first or second column;
- To invest (*I*) in EDP if company 2 chooses the first or third column;
- To invest (*I*) in EDP if company 2 chooses the second or third column;
- To invest (*I*) in EDP if company 2 chooses the first or fourth column;
- To invest (*I*) in EDP if company 2 chooses the second or fourth column;
- To invest (*I*) in EDP if company 2 chooses the third or fourth column;
- To not invest (NI) in EDP, unless company 2 chooses the first column;
- To not invest (NI) in EDP, unless company 2 chooses the second column;
- To not invest (NI) in EDP, unless company 2 chooses the third column;

- To not invest (NI) in EDP, unless company 2 chooses the fourth column;
- To not invest (NI) in EDP no matter which of the four columns company 2 chooses;

Hence, a sixteen-by-four costs matrix can be constructed for the second-level multi-plant safety and security meta-game. This matrix is displayed in Figure 9.4. However, it should be noted that the multi-plant S&S strategy to be developed in a real industrial setting depends on many safety and security risk-influencing (technical and organizational) factors. The table showed in Figure 9.4 can thus very probably be reduced in practical applications.

This second-level meta-game has the same paradox equilibrium point (row 16, column 2), but also another two equilibrium points: (row 7, column 3) and (row 14, column 3). These are equilibrium points because neither company has cause to regret its choice of strategy. For this second-level multi-plant security meta-game, the paradox equilibrium point is clearly not the solution, since a better pay-off can be obtained for both companies from either of the other two equilibrium points. Of the other two equilibrium points, row 14 dominates row 7, and so row 7 is inadmissible as a strategy for company 1.

We can thus conclude that the solution is the following. Company 1 should choose to invest (*I*) in external domino-prevention measures only if it expects company 2 to choose the same strategy as company 1 is expected to make. Otherwise, company 1 should choose to not invest (NI). Company 2 should choose the same strategy as company 1 is expected to choose.

It is obvious that the outcome of the game depends on each company being able to predict what the other, adjacent, company will elect to do. Of course, such predictions are usually impossible in real-life industrial settings. Moreover, the results of this exercise indicate that it might be important for companies to conceal their true intentions or to misrepresent them deliberately.

Using the outcome of the external domino-effects game, a recommendation for helping companies obtaining the socio-economic optimal pay-offs can be made. Assessing the probabilities and/or consequences of external domino-effect risks may lead to the conclusion that it is acceptable to tolerate such risks. Decisions are likely to be based on the perception that the occurrence probability of such risks is so low that they can be ignored. Basically, this means that companies will most probably do whatever is necessary to recover from a major multi-company accident when it occurs, but it will not do all that is necessary to prepare for it in advance. However, the possibility certainly exists that the devastating effects of an induced major accident are so extreme that it is not possible to restore production and/or storage.

The game-theoretical nature of multi-plant safety and security between neighboring plants suggests that neighboring-plant cooperation needs to be dealt with by a supra-plant approach, such as explained in Chapter 4. Chapter 4 discusses the elaboration of a so-called multi-plant council to obtain continuous safety and security improvement within chemical industrial areas, based on steered collaboration. This multi-plant council should have the accepted authority within a

Company 2 / Company 1	*I*, *regardless of company 1*	*NI*, *regardless of company 1*	*Choose same strategy as company 1*	*Choose opposite strategy to company 1*
I, *no matter which column company 2 chooses*	$c_1^{2I}; c_2^{1I}$	$c_1^{2NI} + P_{2,1}L_1; P_{2,2}L_2$	$c_1^{2I}; c_2^{1I}$	$c_1^{2NI} + P_{2,1}L_1; P_{2,2}L_2$
I, *unless company 2 chooses column 4*	$c_1^{2I}; c_2^{1I}$	$c_1^{2NI} + P_{2,1}L_1; P_{2,2}L_2$	$c_1^{2I}; c_2^{1I}$	$P_{1,1}L_1; c_2^{1NI} + P_{1,2}L_2$
I, *unless company 2 chooses column 3*	$c_1^{2I}; c_2^{1I}$	$c_1^{2NI} + P_{2,1}L_1; P_{2,2}L_2$	$P_{1,1}L_1(1-P_{2,1}) + P_{2,1}L_1(1-P_{1,1}); P_{2,2}L_2(1-P_{1,2}) + P_{1,2}L_2(1-P_{2,2})$	$c_1^{2NI} + P_{2,1}L_1; P_{2,2}L_2$
I, *unless company 2 chooses column 2*	$c_1^{2I}; c_2^{1I}$	$P_{1,1}L_1(1-P_{2,1}) + P_{2,1}L_1(1-P_{1,1}); P_{2,2}L_2(1-P_{1,2}) + P_{1,2}L_2(1-P_{2,2})$	$c_1^{2I}; c_2^{1I}$	$c_1^{2NI} + P_{2,1}L_1; P_{2,2}L_2$
I, *unless company 2 chooses column 1*	$P_{1,1}L_1; c_2^{1NI} + P_{1,2}L_2$	$c_1^{2NI} + P_{2,1}L_1; P_{2,2}L_2$	$c_1^{2I}; c_2^{1I}$	$c_1^{2NI} + P_{2,1}L_1; P_{2,2}L_2$
I, *if company 2 chooses column 1 or 2*	$c_1^{2I}; c_2^{1I}$	$c_1^{2NI} + P_{2,1}L_1; P_{2,2}L_2$	$P_{1,1}L_1(1-P_{2,1}) + P_{2,1}L_1(1-P_{1,1}); P_{2,2}L_2(1-P_{1,2}) + P_{1,2}L_2(1-P_{2,2})$	$P_{1,1}L_1; c_2^{1NI} + P_{1,2}L_2$
I, *if company 2 chooses column 1 or 3*	$c_1^{2I}; c_2^{1I}$	$P_{1,1}L_1(1-P_{2,1}) + P_{2,1}L_1(1-P_{1,1}); P_{2,2}L_2(1-P_{1,2}) + P_{1,2}L_2(1-P_{2,2})$	$c_1^{2I}; c_2^{1I}$	$P_{1,1}L_1; c_2^{1NI} + P_{1,2}L_2$
I, *if company 2 chooses column 2 or 3*	$P_{1,1}L_1; c_2^{1NI} + P_{1,2}L_2$	$c_1^{2NI} + P_{2,1}L_1; P_{2,2}L_2$	$c_1^{2I}; c_2^{1I}$	$P_{1,1}L_1; c_2^{1NI} + P_{1,2}L_2$
I, *if company 2 chooses column 1 or 4*	$c_1^{2I}; c_2^{1I}$	$P_{1,1}L_1(1-P_{2,1}) + P_{2,1}L_1(1-P_{1,1}); P_{2,2}L_2(1-P_{1,2}) + P_{1,2}L_2(1-P_{2,2})$	$P_{1,1}L_1(1-P_{2,1}) + P_{2,1}L_1(1-P_{1,1}); P_{2,2}L_2(1-P_{1,2}) + P_{1,2}L_2(1-P_{2,2})$	$c_1^{2NI} + P_{2,1}L_1; P_{2,2}L_2$
I, *if company 2 chooses column 2 or 4*	$P_{1,1}L_1; c_2^{1NI} + P_{1,2}L_2$	$c_1^{2NI} + P_{2,1}L_1; P_{2,2}L_2$	$P_{1,1}L_1(1-P_{2,1}) + P_{2,1}L_1(1-P_{1,1}); P_{2,2}L_2(1-P_{1,2}) + P_{1,2}L_2(1-P_{2,2})$	$c_1^{2NI} + P_{2,1}L_1; P_{2,2}L_2$
I, *if company 2 chooses column 3 or 4*	$P_{1,1}L_1; c_2^{1NI} + P_{1,2}L_2$	$P_{1,1}L_1(1-P_{2,1}) + P_{2,1}L_1(1-P_{1,1}); P_{2,2}L_2(1-P_{1,2}) + P_{1,2}L_2(1-P_{2,2})$	$c_1^{2I}; c_2^{1I}$	$c_1^{2NI} + P_{2,1}L_1; P_{2,2}L_2$
NI, *unless company 2 chooses column 1*	$c_1^{2I}; c_2^{1I}$	$P_{1,1}L_1(1-P_{2,1}) + P_{2,1}L_1(1-P_{1,1}); P_{2,2}L_2(1-P_{1,2}) + P_{1,2}L_2(1-P_{2,2})$	$P_{1,1}L_1(1-P_{2,1}) + P_{2,1}L_1(1-P_{1,1}); P_{2,2}L_2(1-P_{1,2}) + P_{1,2}L_2(1-P_{2,2})$	$P_{1,1}L_1; c_2^{1NI} + P_{1,2}L_2$
NI, *unless company 2 chooses column 2*	$P_{1,1}L_1; c_2^{1NI} + P_{1,2}L_2$	$c_1^{2NI} + P_{2,1}L_1; P_{2,2}L_2$	$P_{1,1}L_1(1-P_{2,1}) + P_{2,1}L_1(1-P_{1,1}); P_{2,2}L_2(1-P_{1,2}) + P_{1,2}L_2(1-P_{2,2})$	$P_{1,1}L_1; c_2^{1NI} + P_{1,2}L_2$
NI, *unless company 2 chooses column 3*	$P_{1,1}L_1; c_2^{1NI} + P_{1,2}L_2$	$P_{1,1}L_1(1-P_{2,1}) + P_{2,1}L_1(1-P_{1,1}); P_{2,2}L_2(1-P_{1,2}) + P_{1,2}L_2(1-P_{2,2})$	$c_1^{2I}; c_2^{1I}$	$P_{1,1}L_1; c_2^{1NI} + P_{1,2}L_2$
NI, *unless company 2 chooses column 4*	$P_{1,1}L_1; c_2^{1NI} + P_{1,2}L_2$	$P_{1,1}L_1(1-P_{2,1}) + P_{2,1}L_1(1-P_{1,1}); P_{2,2}L_2(1-P_{1,2}) + P_{1,2}L_2(1-P_{2,2})$	$P_{1,1}L_1(1-P_{2,1}) + P_{2,1}L_1(1-P_{1,1}); P_{2,2}L_2(1-P_{1,2}) + P_{1,2}L_2(1-P_{2,2})$	$c_1^{2NI} + P_{2,1}L_1; P_{2,2}L_2$
NI, *no matter which column company 2 chooses*	$P_{1,1}L_1; c_2^{1NI} + P_{1,2}L_2$	$P_{1,1}L_1(1-P_{2,1}) + P_{2,1}L_1(1-P_{1,1}); P_{2,2}L_2(1-P_{1,2}) + P_{1,2}L_2(1-P_{2,2})$	$P_{1,1}L_1(1-P_{2,1}) + P_{2,1}L_1(1-P_{1,1}); P_{2,2}L_2(1-P_{1,2}) + P_{1,2}L_2(1-P_{2,2})$	$P_{1,1}L_1; c_2^{1NI} + P_{1,2}L_2$

Figure 9.4 Costs matrix of companies 1 and 2 for the second-level multi-plant S&S meta-game.

chemical multi-plant area to help adjacent companies making collateral agreements on investing in multi-plant S&S prevention and in this way achieving the socio-economic optimum.

9.5
Multi-Plant Games

Concerning external domino risks, inefficient, insufficient, inadequate or even ineffective decisions of one company in a chemical cluster can have devastating impacts on other companies of the cluster. These negative externalities are an important characteristic in what was called the multi-plant domino-effect game (MPDE game). As already mentioned, the return to a chemical plant's investments in domino-effects prevention depends on the domino-effects prevention actions of other chemical plants from the multi-plant area in which the plant is situated. If other companies do invest, then investing is more attractive, and if they do not, then it is less interesting. Hence, the MPDE game exhibits "strategic complementarities" (Milgrom and Roberts, 1990).

It should be noted that the more that companies invest in domino-effects prevention, the lower are the negative externalities in the system. This game can have one or multiple Nash equilibria. For the case where there are two Nash equilibria – either all companies invest in domino-effects prevention or none of the companies do – the possibility of so-called "tipping" exists, indicating that inducing some companies to invest in domino-effects prevention will lead others to do so as well. Therefore, the possibility of tipping the Nash equilibria from a state of "no investment" to one of "universal investment" (by all companies) as regards domino-effects prevention in a multi-plant area is investigated hereafter.

Multi-person games of strategy are games involving three or more players, each of whom has partial control over the outcome. It is obvious that such games potentially involve coalitions. The MPDE game is a partially cooperative multi-person game in which coalition forming is allowed and even essential. Let a "critical coalition" be a coalition of chemical companies where a change from "not investing in domino-effects prevention" to "investing in domino-effects prevention" by its members will induce all non-members of the coalition to follow suit.

Consider x chemical companies composing a chemical cluster $\{x\}$. Let the companies be indexed by i. Every company is characterized by (i) the probability P_i that company i's actions (or lack of actions) lead to a direct loss L_i (caused by internal domino effects), and (ii) the investment in domino-effects prevention at a cost c_i that leads for company i to avoidance of direct loss with certainty (it should be noted that c_i can thus be interpreted as a "hypothetical benefit" of avoiding domino effects by investments in prevention). If company i incurs a direct loss, then this may also affect other companies' outcomes. If company i does not incur a direct loss then it will have no negative impact on other companies. The loss to other companies (caused by external domino effects) is considered as "an indirect impact". Let $l_i(\{y\}, S_i)$ then be the expected indirect loss to chemical plant i when it follows strategy S_i and the companies in the chemical sub-cluster $\{y\}$ are the only ones from the chemical cluster $\{x\}$ investing in domino-effects prevention ($y \le x$). It should be noted that the model is conceptualized such that a company who has invested in domino-effects prevention cannot cause an indirect impact on others.

Furthermore, if every other company than company i invests in domino-effects prevention, then company i cannot suffer indirect impacts (i.e. impacts from other companies). In other words, if $\{y\} = \{1, 2, ..., i-1, i+1, ..., x\}$ then $l_i(\{y\}, S_i) = 0$, independent of the situation where i invests or does not invest in domino-effects prevention.

Assume that company i invests in domino-effects prevention. Furthermore, assume that chemical companies belonging to the sub-cluster $\{y\}$ also invest in domino-effects prevention. Then, the expected loss to company i is $c_i + l_i(\{y\}, I)$, whereby the second term in this formula is the expected cost of indirect impacts (consequences) imposed by companies belonging to $\{x\}$ who do not invest in domino-effects prevention. The expected loss of company i not investing in domino-effects prevention can be expressed as:

$$P_i L_i \prod_{j \neq i, j \in \{x/y\}} (1-P_j) + (1-P_i) \cdot l_i(\{y\}, \mathrm{NI}) \tag{9.6}$$

Hence, direct (internal) domino effects (the first term in the latter expression) are conditioned on the non-occurrence of indirect losses. Indirect effects (the second term in the latter expression) are conditioned on direct losses not occurring. These conditions result from the fact that a chemical installation can only explode or be destroyed once or that internal and external domino effects do not originate at the same time.

A chemical company i is indifferent between investing and not investing in domino-effects prevention if:

$$c_i + l_i(\{y\}, I) = P_i L_i \prod_{j \neq i, j \in \{x/y\}} (1-P_j) + (1-P_i) \cdot l_i(\{y\}, \mathrm{NI}) \tag{9.7}$$

Hence, the cost of investment in domino-effects prevention at which company i is indifferent, can be defined:

$$\tilde{c}_i(\{y\}) = P_i L_i \prod_{j \neq i, j \in \{x/y\}} (1-P_j) + (1-P_i) \cdot l_i(\{y\}, \mathrm{NI}) - l_i(\{y\}, I) \tag{9.8}$$

If $c_i > \tilde{c}_i(\{y\})$, then company i will not invest in domino-effects prevention, and *vice versa*.

Furthermore, if a chemical company within a chemical cluster $\{x\}$ decides to invest in domino-effects prevention, the sub-cluster $\{y\}$ is increased by one unit. As a result, the expected indirect loss, $P_i l_i(\{y\}, \mathrm{NI})$, decreases. Following, $\tilde{c}_i(\{y\})$ increases in $\{y\}$, since more chemical companies investing in domino-effects prevention leads to lower expected indirect losses. Thus, the maximum cost at which domino-effects-prevention investments are justified, increases with every company deciding to invest in such prevention.

A Nash equilibrium for the MPDE game is a set of pure strategies $S_1, \ldots, S_x$ such that (i) $S_i = I, \forall i \in \{y\}$ (which may be empty), (ii) if $\tilde{c}_i(\{y\}) > c_i$ then $S_i = I$, (iii) if $\tilde{c}_i(\{y\}) < c_i$ then $S_i = \text{NI}$, and (iv) company i is indifferent between I and NI, then $\tilde{c}_i(\{y\}) = c_i$.

There may be equilibria where all chemical companies invest in domino-effects prevention, those where none do, and asymmetric pure strategy equilibria where some plants invest and others do not. If there are two equilibria, one with all enterprises not investing and the other with everyone investing in domino-effects prevention, then it should be investigated as to how to possibly tip the socially sub-optimal equilibrium (NI, NI, ..., NI) to a socially optimal equilibrium ($I, I, \ldots, I$). Hence, it is examined how the MPDE game with two (or more) equilibria may be tipped by a change in the strategy choices of a small number of players.

Let (NI, NI, ..., NI) be a Nash equilibrium. A tipping-inducing sub-cluster (TISC) for this equilibrium is a set $\{z\}$ of chemical companies such that if $S_i = I, \forall i \in \{z\}$, then $\tilde{c}_j(\{z\}) \geq c_j, \forall j \notin \{z\}$. In other words, a tipping-inducing sub-cluster is a sub-cluster with the property that if all chemical plants belonging to that sub-cluster decide to invest in domino-effects prevention, then for all other companies belonging to the entire chemical cluster the best strategy is also to invest in such prevention. A "minimum TISC" is a TISC of which no sub-set is also a TISC, indicating that all companies in the minimum TISC are required to tip the other (non-investing) companies into a domino-effects investment strategy. Furthermore, since there can be several minimum TISCs and we are only interested in the one containing the smallest number of companies, we let the "smallest number minimum TISC" be a minimum TISC where no other TISC includes fewer companies. Assume that the change in the expected indirect loss to plant i, who does not invest in domino-effects prevention, when company j joins the set $\{y\}$ of companies who have already invested in domino-effects prevention, is:

$$l_i^j(\{y\}, \text{NI}) = l_i(\{y/j\}, \text{NI}) - l_i(\{y\}, \text{NI}) \geq 0, \forall j \in \{y\} \tag{9.9}$$

Heal and Kunreuther (2007) then prove that a smallest number minimum TISC is easily characterized. They further indicate (in general terms) that an equilibrium where no company invests (e.g., in domino-effects prevention) may be converted to one with full investment by persuading a sub-set of the companies to change their policies. Hence, the least expensive way of changing equilibrium is providing incentives for a tipping-inducing sub-cluster to change its behavior, which will then tip the entire cluster.

Hence, a TISC does exist in theory. In order to help clear understanding, an illustrative example is provided of how such tipping might occur in a chemical cluster consisting of three companies. First, the illustrative example is given in general terms and at the end, numerical values are used to show the DE game's validity and its potential in a real industrial setting.

For simplicity, consider a cluster of three companies 1, 2 and 3. Investigation is conducted as to whether it is possible, and under what conditions, for one company by changing its strategy from NI to I, to tip the other two companies to change their strategies as well from NI to I.

Company 2 \ Company 3	*I*	*NI*
I	c_2^{3I} ; c_3^{2I}	$c_2^{3NI} + P_{3,2}L_2$; $P_{3,3}L_3$
NI	$P_{2,2}L_2$;$c_3^{2NI} + P_{2,3}L_3$	$P_{2,2}L_2(1-P_{3,2}) + P_{3,2}L_2(1-P_{2,2})$; $P_{3,3}L_3(1-P_{2,3}) + P_{2,3}L_3(1-P_{3,3})$

Figure 9.5 Costs matrix of companies 2 and 3 in the case where the strategy of company 1 is *I*.

To investigate whether it is possible in the three-company case study for one company to tip the other two companies into changing their strategies, the existing Nash equilibria need to be determined, and the conditions under which strategies are dominant have to be established. Therefore, the costs matrices for two possible cases are included: (i) the strategy of company 1 is "*I*", and (ii) the strategy of company 1 is "NI".

The case where the strategy of company 1 is "*I*" is first considered. If all three companies invest in domino prevention, then their costs are just their investment costs, c_i. If (besides company 1) company 2 invests in domino-effects prevention, and company 3 does not invest, then companies 1 and 2 incur their investment cost c_i plus the expected loss from a domino effect from company 3 onto, respectively, company 1 or company 2 (i.e. $P_{3,1}L_1$, respectively, $P_{3,2}L_2$). Company 3 just has an expected loss from an internal domino effect, that is $P_{3,3}L_3$. If neither company 2 nor company 3 invest, then company 1 has an expected loss of its investment costs plus the expected loss from a domino effect from company 2 onto company 1 (i.e. $P_{2,1}L_1$), conditioned on there being no domino effect from company 3 onto company 1 (i.e. times $1 - P_{3,1}$), plus the expected loss from a domino effect from company 3 onto company 1 (i.e. $P_{3,1}L_1$), conditioned on there being no domino effect from company 2 onto company 1 (i.e. times $1 - P_{2,1}$). All the other expected losses composing the costs matrix are determined in a similar way. The resulting costs matrix can be found in Figure 9.5.

When company 1's strategy is to invest (*I*), choosing *I* is a dominant strategy for company 2 under the conditions that

$$\begin{cases} c_2^{3I} < P_{2,2}L_2 \\ c_2^{3NI} + P_{3,2}L_2 < P_{2,2}L_2(1-P_{3,2}) + P_{3,2}L_2(1-P_{2,2}) \end{cases} \tag{9.10}$$

Or:

$$\begin{cases} c_2^{3I} < P_{2,2}L_2 \\ c_2^{3NI} < P_{2,2}L_2(1-2P_{3,2}) \end{cases} \tag{9.11}$$

Company 3 / Company 2	*I*	*NI*
I	$c_2^{3NI} + P_{1,2}L_2 ; c_3^{2I} + P_{3,3}L_3$	$c_3^{2NI} + P_{1,2}L_2(1-P_{3,2}) + P_{3,2}L_2(1-P_{1,2})$; $P_{3,3}L_3(1-P_{1,3}) + P_{1,3}L_3(1-P_{3,3})$;
NI	$P_{2,2}L_2(1-P_{1,2}) + P_{1,2}L_2(1-P_{2,2})$; $c_3^{2NI} + P_{1,3}L_3(1-P_{2,3}) + P_{2,3}L_3(1-P_{1,3})$	$P_{2,2}L_2(1-P_{1,2})(1-P_{3,2}) + P_{1,2}L_2(1-P_{2,2})(1-P_{3,2}) + P_{2,2}L_2(1-P_{1,2})(1-P_{2,2})$; $P_{3,3}L_3(1-P_{1,3})(1-P_{2,3}) + P_{1,3}L_3(1-P_{3,3})(1-P_{2,3}) + P_{2,3}L_3(1-P_{3,3})(1-P_{1,3})$

Figure 9.6 Costs matrix of companies 2 and 3 in the case where the strategy of company 1 is NI.

The first condition is obviously what we would expect to be true in the case of a single chemical company deciding whether to invest in domino-effects prevention or not, thereby not taking into account the strategies of the other two companies within the cluster of three companies. The second condition, expressing the domino-effect risk from a nearby company (thus considering the existence of the cluster in which the company is situated), is evidently stricter than the first condition.

Furthermore, (I, I) is a Nash equilibrium if $c_i < P_{i,i}L_i$ and is a dominant strategy if $c_i < P_{i,i}L_i(1-2P_{j,i})$ with $i = 2$ or 3. (NI, NI) is a Nash equilibrium if $c_i > P_{i,i}L_i(1-2P_{j,i})$ and is a dominant strategy if $c_i > P_{i,i}L_i$.

For the case where company 1 does not invest (strategy NI), a matrix representing the costs incurred by companies 2 and 3 may be determined as well. Figure 9.6 illustrates this matrix.

It should be noted that the further chain of indirect domino losses, created by a sequence of breakdowns and characterized by multiplied failure probabilities (e.g., $P_{1,2}P_{2,3}$ or $P_{2,1}P_{1,3}$), is not taken into account in the model simply because the multiplication results are negligible.

In the case company 1's strategy is to not invest (NI), choosing I is a dominant strategy for company 2 under the conditions that

$$\begin{cases} c_2 + P_{1,2}L_2 < P_{2,2}L_2(1-P_{1,2}) + P_{1,2}L_2(1-P_{2,2}) \\ c_2 + P_{1,2}L_2(1-P_{3,2}) + P_{3,2}L_2(1-P_{1,2}) < P_{2,2}L_2(1-P_{1,2})(1-P_{3,2}) + P_{1,2}L_2(1-P_{2,2})(1-P_{3,2}) \\ \qquad + P_{3,2}L_2(1-P_{1,2})(1-P_{2,2}) \end{cases} \tag{9.12}$$

The conditions for which I is a dominant strategy for company 3 can be derived analogously. Furthermore, (I, I) is a Nash equilibrium if $c_i + P_{j,i}L_i < P_{i,i}L_i(1-P_{j,i}) + P_{j,i}L_i(1-P_{i,i})$ with $(i, j) = (2, 1)$ and $(i, j) = (3, 1)$.

Our MPDE game is thus characterized by multiple Nash equilibria. Hence, if both cost matrices from company 1 using "*I*" (Figure 9.5) or "NI" (Figure 9.6) as a strategy are considered, the required conditions to turn the decision of companies 2 and 3 from "not investing in domino-prevention measures" (being part of a Nash equilibrium, which is obviously not optimal from a socio-economic viewpoint), to "investing in domino-prevention measures" (being part of a socio-economic optimal Nash equilibrium) can be determined.

The tipping problem is illustrated using the following numerical example. Let $c_1 = 4000€$, $c_2 = c_3 = 700€$. Assume that the probabilities aggregated for all installations within a company and per 10000 years, are $P_{2,2} = P_{2,3} = P_{2,1} = P_{3,2} = P_{3,3} = P_{3,1} = 0.1$; $P_{1,1} = 0.2$; $P_{1,2} = P_{1,3} = 0.3$. Assume further that the potential company losses are $L_1 = 20000€$ and $L_2 = L_3 = 10000€$.

It is examined what the dominant strategy is for company 1, given that companies 2 and 3 (considered in a sub-cluster of companies) are currently not investing in domino prevention ($\{y\} = \phi$) and do not consider the losses possibly resulting from the other companies in the cluster. In that case the direct losses to companies 2 and 3 are $c_2(\phi) = c_3(\phi) = P_{2,2}L_2(1 - P_{1,2})(1 - P_{3,2}) = 600€$. It is obvious that, since $c_2 > c_2(\phi)$ and $c_3 > c_3(\phi)$, neither company 2 nor company 3 will invest in domino-effects prevention if company 1's strategy is NI. Since $c_1(\varphi) = P_{1,1}L_1(1 - P_{3,1})(1 - P_{2,1}) = 3200€$ is smaller than c_1 (= 4000€), company 1 will indeed not invest and (NI, NI, NI) is actually a Nash equilibrium. Hence, if company 1 does not invest, then not investing is a dominant strategy for the other companies for all $c_i > 600€$.

Given that the three companies are located within the same cluster, limiting the analysis to direct (internal) domino effects does not offer a solid basis for domino-prevention management. Also, the indirect loss caused by the fact that only a sub-set $\{y\}$ of companies is investing in prevention should be considered. An example of how including the indirect risks can alter the analysis, is therefore provided. Assume that company 1 is obliged to invest in domino-effects prevention (e.g., due to national or regional regulations and/or legislation) and as a result no negative externality from company 1 is imposed on the other companies (2 and 3). If this is the case, the indirect risk to chemical company i (2 or 3) is the same whether company i (2 or 3) itself decides to "invest" or decides to "not invest", that is $l_i(\{y\}, NI) = l_i(\{y\}, I)$ (where $i = 2, 3$), hence expression

$$\tilde{c}_i(\{y\}) = P_i \cdot \left(L_i \prod_{j \neq i, j \in \{x/y\}} (1 - P_j) - l_i(\{y\}, \mathrm{NI}) \right) \tag{9.13}$$

can be used to determine the critical cost levels of companies 2 and 3, which in both cases amount to 800€.

Since both c_2 and c_3 are smaller, companies 2 and 3 will be changing their strategy of NI to *I* as a result of company 1 doing so. Therefore, company 1 has the power to tip the other companies within the three-company cluster from one strategy to another strategy. In other words, one company's strategic choice concerning domino prevention may significantly influence the other companies' domino-prevention-related strategic decisions in a chemical cluster. This is an important finding implying that, for example, company-specific incentives or well-elaborated domino-prevention regulations can lead to substantial safety and security improvements within chemical clusters.

The obvious advantages of the multi-plant game-theoretic approach elaborated in this section may lead to improving safety and security in multi-plant areas by possibly increasing the investments in external domino-effects prevention in such settings.

9.6 Summary and Conclusions

In most chemical companies nowadays, safety and security investments are made for two main reasons: to meet the minimum legal requirements and to significantly lessen the risk of internal accidents. Chemical plants are, however, located almost always in clusters of two or more companies. Therefore, they can also be affected by accidents triggered by nearby plants. Company top management should realize that their neighbours investing in external domino-effects prevention may actually avoid a major accident happening within their own company. Top management should also realize that joint strategic cross-plant investments by different plants within a chemical cluster may be far more cost-efficient than plants taking individual external domino-effect-prevention investments.

This chapter offers new insights from a game-theoretical viewpoint into tackling the very complex decision problem of cross-plant escalation effects. This research examined the extent to which game theory is applicable to safety and security policy within a two-company chemical cluster. From the moment there is some degree of interaction between the neighboring companies, investment decisions can be described as a strategic game. The most effective method to persuade companies to invest in external domino-effect prevention is either to establish a multi-plant initiative by several plants belonging to a chemical cluster, or introducing taxes or awarding subsidies by the government, or raising or lowering the insurance premiums by insurance companies.

Stability in the strategic decisions of adjacent collaborating companies is to be expected either when joint investments in the prevention of external domino effects can take place at a sufficiently low cost or when deliberate incentives are provided.

To determine the steps that need to be taken in a chemical multi-plant area for socio-economically optimizing the existing situation as regards domino-effects prevention, an investigation is required to identify how single chemical plants manage their external domino-effect risks where there is a likelihood that even if they have decided to invest in domino-prevention measures, they might be harmed due to other companies not following suit.

Players (adjacent chemical companies) in the multi-plant safety and security game may end up playing a socio-economic sub-optimal Nash equilibrium (i.e. a win-win situation of strategic decisions) instead of the unstable collective self-interest equilibrium (i.e. a win-win situation of strategic decisions with higher pay-offs, or lower expected losses). The study described in Section 9.5 was carried out from the perspective that academic research can help such companies identify the appealing possibilities at their disposal to reach an agreement on playing such collective self-interest equilibrium or in other words an optimizing martyrdom equilibrium point. If recommendations can be made for changing the strategic choice of two adjacent companies from not investing to investing in external domino-effects prevention, a socio-economic non-optimal situation can be changed to a socio-economic optimal situation.

For companies to take socio-economic optimal strategic decisions/positions, using the game-theoretical approach, authorities can calculate subsidies and/or taxes for giving incentives. If academic research would further lead within chemical clusters to the setting up of a multi-plant institution at cluster level with the ability to persuade neighboring plants into changing their strategic positions (by increasing trust, making unambiguous agreements, *etc.*), safety and security might truly be optimized within a chemical industrial area.

10
Conclusions and Recommendations

10.1
Introduction

In general, chemical engineers in industry encounter a variety of complex chemical processes. These processes often involve substances of high chemical reactivity, high toxicity, and high corrosivity operating at high temperatures and pressures. These characteristics can lead to a variety of potentially serious consequences, including explosions, environmental damage, and threats to people's health. It is essential that errors or omissions resulting from missed communication between persons and/or groups involved in design and operation do not occur when dealing with chemical processes. It is also imperative that intentional attacks on critical infrastructure are identified and prevented. The latter concerns apply to both single-plant personnel and personnel belonging to a cluster of plants.

This book serves as a guide to elaborating and determining the principles, assumptions, strengths, limitations and application areas of cross-company chemical safety and security management. The document offers instruments, guidelines, procedures, and frameworks for creating a safety and security culture in a cluster of chemical enterprises. The presentation is conceptually rather than mathematically oriented to maximize the application potential "in the field", that is, in the chemical industry. The author hopes that plant safety and security managers, regulators, lobbyists and other readers will acquire a perception of the optimization possibilities inherently present in enhanced chemical safety and security management on an aggregated multi-plant level. However, the aim of this document surpasses the intended support of multi-plant safety and security awareness and offers, as already mentioned, easy-to-use information for actually setting up a multi-plant-safety and -security culture.

10.2
Summary

This book explores the concept of "multi-plant safety and security" in the chemical industry. The conceptualization of such aggregated safety and security learning

Multi-Plant Safety and Security Management in the Chemical and Process Industries. G.L.L. Reniers

ISBN: 978-3-527-32551-1

and integrated safety and security knowledge is set to become one of the main challenges of future safety-management optimization as well as security-management improvement in the chemical industry. Elaborating the notion of cross-company safety and security management requires a number of components that have to be thoroughly worked out and efficiently managed, especially for highly diversified and complex surroundings. Therefore, this work gradually develops a set of concepts, procedures and tools needed for modeling a system to extend single-plant safety and-single plant security to multi-plant safety and multi-plant security.

The first chapter examines the definitions, concepts, strategies and characteristics concerning risks associated with processing, storing and handling hazardous substances on an industrial scale. The way these risks, whether it be safety risks or security risks, are identified, analyzed and managed is briefly described. One type of risk in particular that is linked with chemical industrial areas is discussed, that is external domino-effect risks. While they are typified by a combination of very high impacts with very low probabilities, external domino-effect risks are very complex phenomena about which relatively little historical accident data are available. As a result, plant safety and security managers only pay limited attention to them, thus undervaluing their significance within the multi-plant area. No specific management guidelines or user-friendly tools are currently available in a multi-plant context to safety experts or to security specialists to prevent major accidents involving external domino effects.

Chapter 3 discusses how a multi-plant-safety and -security culture can be developed. Three building blocks to conceptualize, to elaborate and to refine a multi-plant-safety and -security culture are suggested: (i) to address safety and security management people collaboration, (ii) to draft multi-plant safety-management system guidelines and multi-plant-security-management-program guidelines, and (iii) to develop cross-company tools and technology to tackle external domino effects.

In Chapter 4, recommendations are given about setting up the first building block, that is drafting multi-plant-safety and -security-management procedures.

A multi-plant-safety-management system and a multi-plant-security-management program are conceptualized. The most important topics for aggregated safety and aggregated security are emphasized and suggestions for expanding the subjects towards multi-plant SMS and multi-plant SMP guideline documents are proposed.

Moreover, a plan is put forward to organize multi-plant safety and security management via the establishment of a so-called "multi-plant (safety and security) council" to be composed of two collaborating parts. The first part, including personnel belonging to the different companies of the multi-plant area (within a larger chemical cluster) is proposed to be divided into eight think tanks. Their cross-company meetings together with a knowledge management network should lead to the enhancement of single-plant safety and security as well as multi-plant safety and security. The second part, which is called the "multi-plant council data administration", is split into a safety and a security sub-part. Both sub-parts have to embody company-independent staff handling confidential individual company data and executing independent research surrounding multi-plant-related safety, respectively, multi-plant-related security issues.

In Chapter 5, multi-plant management frames of reference forming a support for enhanced cross-company safety and security cooperation are shaped.

It is suggested that three very widespread risk-analysis techniques, that is hazop analysis, what-if analysis and the risk matrix can be integrated into one framework called hazwim. The latter can be used by joint plant-safety management to analyze their cross-company safety risks or to design their preferential measures to avoid, as much as possible, cross-company accidents. Hazwim is thus designed to be used in the case of an existing cluster of (at least two) neighboring plants.

Concerning security collaboration, Chapter 5 proposes that a distinction be made between indoor security services and outdoor security services. A framework to help the former collaborate across companies to carry out security-vulnerability assessments is put forward. In addition, the creation is suggested of a flexible, efficient and effective network of cross-competent outdoor security guards by means of intensive and carefully balanced training programs.

In addition, an evaluation instrument was elaborated and discussed for use by the MPC or single plant management to determine the necessary security staffing levels in an industrial complex.

Chapter 6 develops conceptual technology and software design code to help the multi-plant council data administration with its task in looking into cross-company accident prevention. Since at present there appears to be a lack of a user-friendly prevention-oriented computer-automated tool specifically aimed at cross-company *safety and security* related accidents, mathematical concepts for elaborating user-friendly external domino software are discussed. Two modules are suggested. The first module can be used to investigate how to divide industrial areas into smaller parts in between which no domino effects are possible. The second module leads to a ranking based on a so-called segment risk factor (SRF). The latter expresses the relative domino-effects danger of any 3-sequence installations in a complex network composed of a variety of installations.

Chapter 7 discusses how chemical multi-plant areas may efficiently assess and optimize their operational manning levels. A framework and a roadmap to assist chemical plants in assessing and evaluating their operational staffing levels is given. Furthermore, a methodology to assist chemical companies in analyzing substantial changes of their existing operational staffing levels is provided. The method can be implemented within a multi-plant area or within a part thereof (i.e. a single company, a business unit or an installation).

Finally, a performance-based methodology to assist a multi-plant area helping its members (i.e. single chemical plants) in their efforts to systematically improve operational manning levels is worked out. By integrating manning-level standards with the plan-do-check-act cycle of continuous improvement and providing a generic action plan for any organization to elaborate an efficient means of improving manning-level performance, operational manning levels can be advanced without having to undergo a costly and inefficient trial-and-error phase.

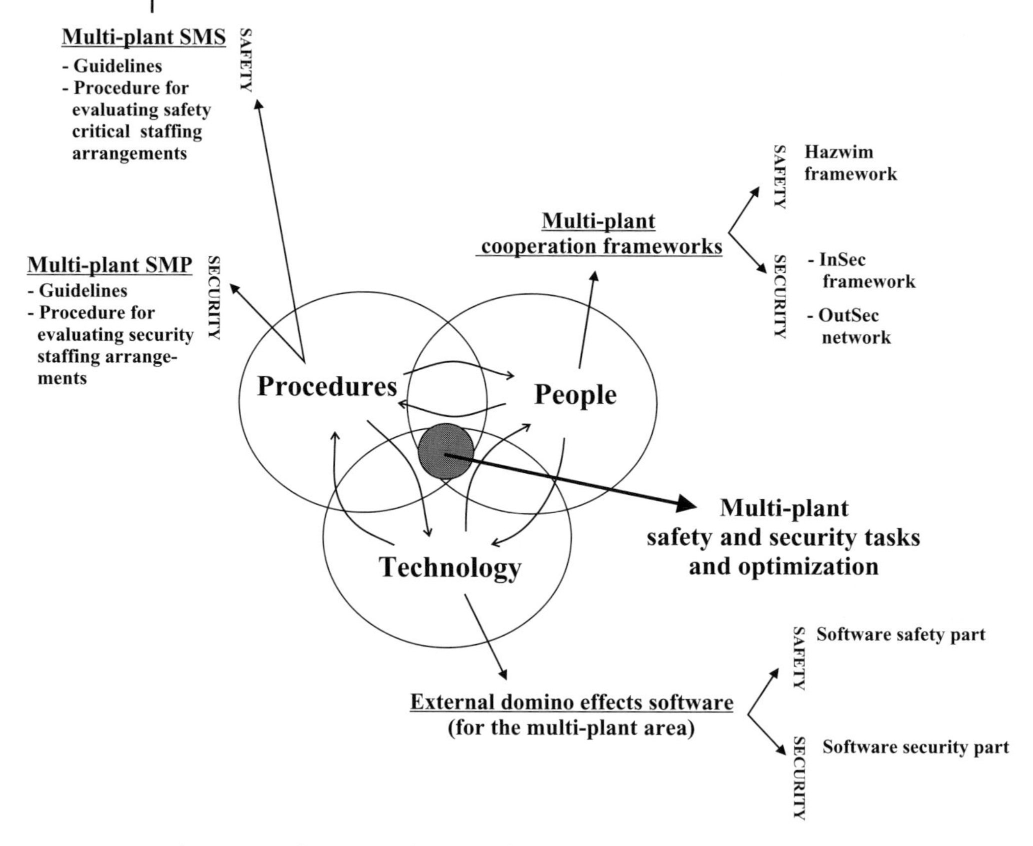

Figure 10.1 Shaping a multi-plant-safety and -security culture in the chemical industry.

As already mentioned, the challenges and obstacles that lie on the path to achieving robust multi-plant safety and security management able to deal with multi-plant safety topics and multi-plant security issues can be classified into three main building blocks: procedures (Chapter 4), people (Chapter 5) and technology (Chapter 6). A conceptual architecture of a multi-plant safety and security culture is proposed in Figure 10.1.

Chapter 8 discusses the integration and the real situation problem of implementing the components of the building blocks proposed in Chapters 4–6. A generally conceptualized management design frame to address the complex elaboration difficulty of the multi-plant safety and security culture is suggested. Furthermore, impulse is given to deal with safety and security within the scope of sustainability. The principles for attaining a sustainable safety and security strategy in chemical clusters are put forward and clarified.

In Chapter 9 game theory is used as a promising scientific technique to investigate the decision-making process on investments in prevention measures simultaneously involving several plants. The game between two neighboring chemical plants and their strategic investment behavior regarding the prevention of external domino effects is described. Recommendations are formulated to advance cross-plant (safety and security) prevention investments in a two-company cluster.

Furthermore, it is investigated whether changing the strategic choice of a small number of companies of a cluster composed of more than two plants, might tip all the rest of the plants within the cluster to change their strategies from a socially non-optimal situation to a socially optimal situation.

The shaping of a cycle for continuously improving safety and security in a cluster of chemical companies and the stages elaborated in this book to this end are illustrated in Figure 10.2a,b, respectively. The drawing is based on a generic model for the development of a management system suggested in ISO Guide 72 (2001).

The generic models illustrated in Figure 10.2a,b are based on the plan-do-check-act principle. Technical prevention of major risks is centrally situated within this Deming loop that subjects the safety and security prevention topics to the demands of all stakeholders concerned. An integrated multi-plant-safety and -security culture is shaped. Figure 10.3 illustrates the theoretical safety and security continuous amelioration picture, especially for preventing major accidents, resulting from implementing the principles given in this book for creating a sustainable safety and security strategy.

The upper half of Figure 10.3 shows the stepwise progress both disciplines, that is safety science and security science, have made over time and are still making. Taking two companies *E* and *F* as an example, every distinct progress step (characterized by new insights) is represented by a rhomb. This rhomb can actually be considered as a kind of "layer of accident prevention" (safety) and/or "ring of accident prevention" (security). The hypothetical development of ten potential external domino accidents for the two plants is shown. Every potential escalation accident is prevented, except for potential multi-plant-accident number eight. External domino accident number five, for example, was stopped by the accident-prevention layer marked by existing safety and security procedures in the multi-plant area including company *E* and company *F*. The lower half of the figure illustrates the chemical industry's ceaseless stepwise sustainability amplification resulting from new insights (captured over time) dealing with safety and security. Each layer of accident prevention is considered as increasing safety and security effectiveness and sustainability.

10.3 Main Conclusions and Recommendations

Seven main conclusions are formulated as recommendations. Every recommendation is introduced by briefly explaining the study leading to the conclusion's innovative character for safety and security management in the chemical industry. A general conclusion is finally set out.

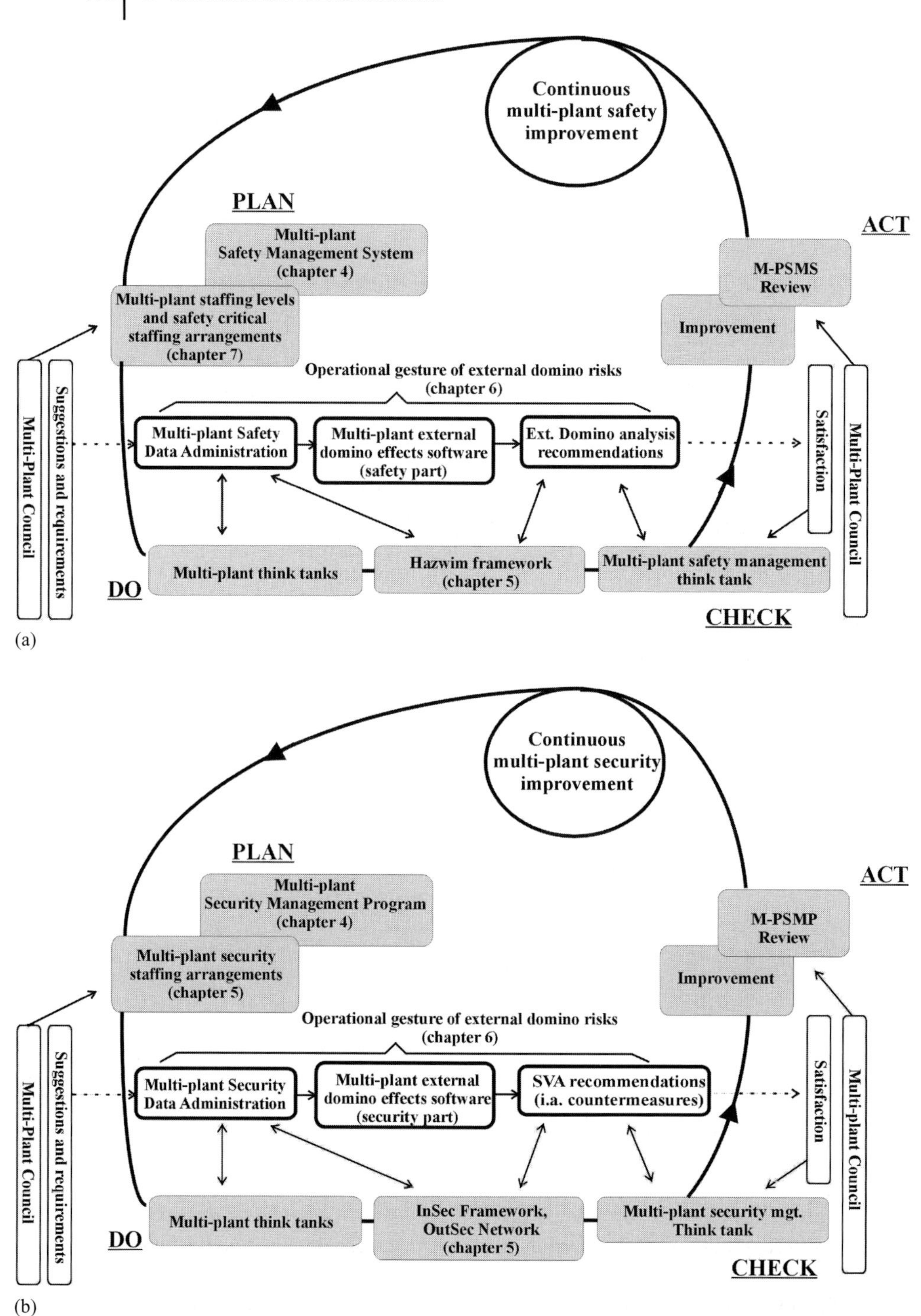

Figure 10.2 (a) Cycle for continuous safety improvement of a chemical industrial area; (b) Cycle for continuous security improvement of a chemical industrial area.

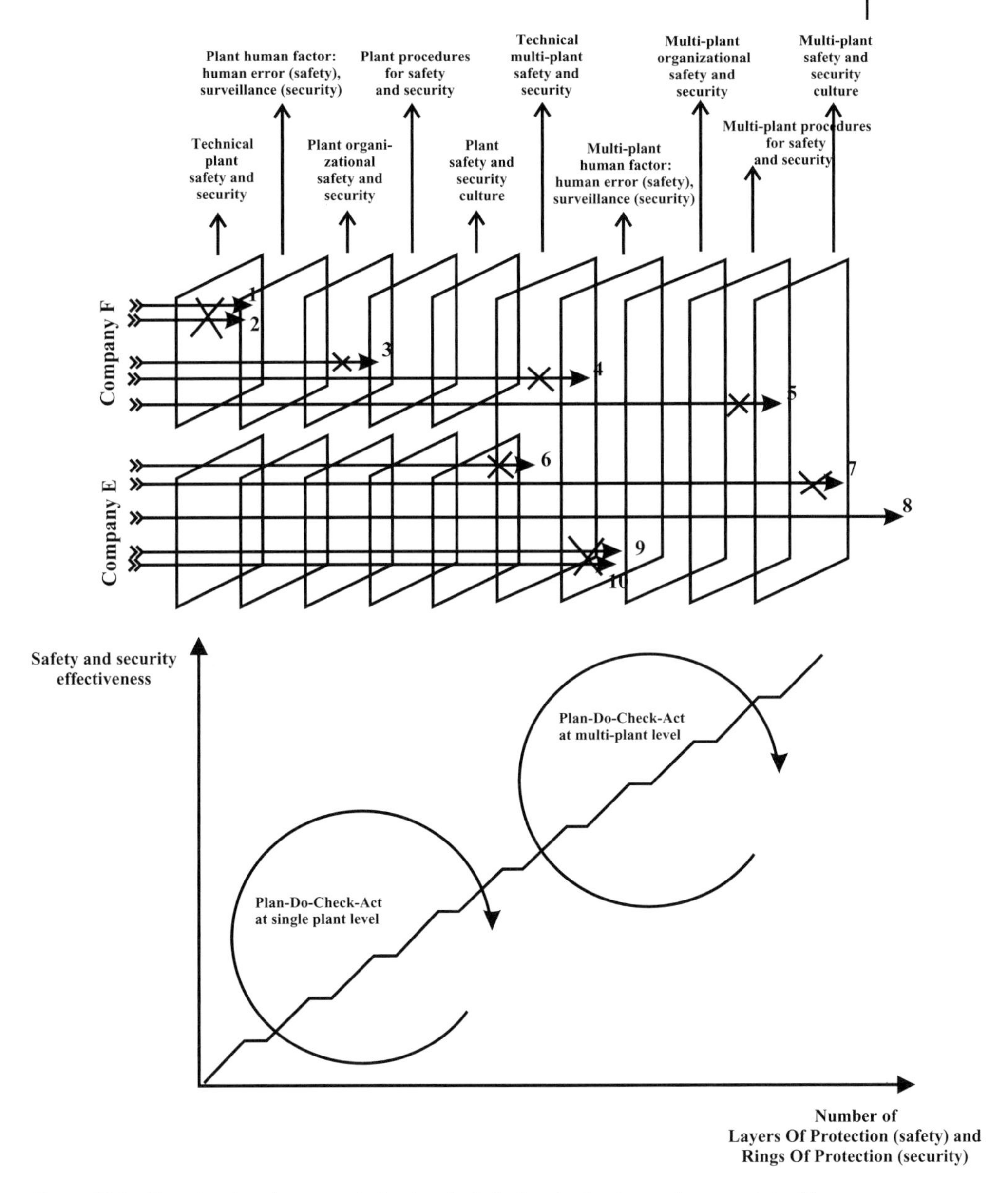

Figure 10.3 Stopping knock-on events in chemical clusters by implementing a sustainable safety and security strategy.

Conclusion 1

Assessing probabilities and consequences of chemical risks may lead to the conclusion that it is acceptable to tolerate a (safety or security) risk. In the case of so-called external domino-effect risks, decisions are likely to be based on the perception that the occurrence probability is so low that it can be ignored.

However, evidence suggests that cross-plant escalation accidents that have actually happened indicate the devastating effects to be so extreme that it is often not possible to restore production and/or storage. Moreover, there is much delegation of major hazard regulatory tasks to the regulated, namely in self-regulation. This is also the case for external domino-accident prevention. The technological nature of cross-company domino effects suggests that multi-plant preventive legislation and multi-plant cooperation both need to be dealt with by an integrated approach. They need to be seen as useful to industry and public sectors by helping these sectors and industries to keep focused on the real cross-company threats they face.

→ *Recommendation 1:*

In the chemical industry, individual plant safety and security cultures need to be upgraded to an integrated multi-plant-safety and -security culture.

Conclusion 2

The change from an individual safety and security culture towards a multi-plant-safety and -security culture would appear to be very difficult to overcome. Several building blocks need to be elaborated and integrated in a multi-plant environment: multi-plant safety and security procedures, multi-plant safety and security collaboration, multi-plant accident-prevention decision-support tools and techniques and multi-plant staffing-level topics. However, all these capabilities already exist within the single plants (although different approaches are used) composing the multi-plant area. Moreover, implementing quality systems and especially the Deming loop of continuous improvement is very widespread.

→ *Recommendation 2:*

The strategy to shape a multi-plant site-integrated safety and security culture involves combining an object-oriented approach (to work out three building blocks) with the loop of continuous improvement (to connect the building blocks and introduce dynamism into the culture).

Conclusion 3

When drafting multi-plant safety and/or security procedures or when planning risk analyses to avoid external domino effects, company confidential information has to be used. Nevertheless, joint safety and security procedures, domino risk identification and safety and security optimization should be based on individual plant information and on collaboration between personnel at different levels of the companies.

→ *Recommendation 3:*

A need exists for an institute at multi-plant level (in this book called the "multi-plant council") responsible for multi-plant safety issues as well as multi-plant security topics and divided into two parts: a part comprising multi-company staff and a part comprising independent personnel handling confidential data.

Conclusion 4

When conceptualizing a workable framework to enhance cross-company collaboration between safety experts or between security specialists, these professionals need to be able to communicate on the same level and with the same knowledge. Using commonly known risk-analysis techniques and vulnerability-assessment procedures serves this purpose.

→ *Recommendation 4:*

Three currently very widespread risk-analysis procedures (Hazop, what-if analysis and the risk matrix) are to be used by cross-company safety management to investigate multi-plant-related safety risks in an existing setting of chemical installations. To facilitate the latter, the elaborated hazwim framework is recommended.

The separation of plant-security services into two parts is recommended: indoor security and outdoor security. In this manner, multi-plant security collaboration can be streamlined using the proposed InSec framework and the suggested OutSec network.

To verify security staffing levels are adequate in a single-plant or a multi-plant environment, a specific instrument should be employed.

Conclusion 5

A user-friendly approach was developed to relatively rank installations in an industrial surrounding, providing multi-plant-safety management and multi-plant-security management with precautions planning information. Instead of focusing on the intrinsic danger of an installation (as programs used in industrial practice usually do), the installations listing is based on the propagation features of an installation.

→ *Recommendation 5:*

A need exists to use user-friendly software for ranking installations with respect to both initiating and continuing domino effects in a chemical cluster. By concentrating on the knock-on factor of domino events rather than on the events themselves, taking precautions in a chemical multi-plant area can be achieved in a more cost-effective way from a managerial point-of-view. By using this innovative approach, it is more realistic in industrial practice to take preventive measures to minimize the spread (and thus the consequences) of (accidental as well as intentional) domino effects.

Conclusion 6

Chemical enterprises and multi-plant areas are often challenged with efficiently assessing and optimizing their operational manning levels. Unbalanced and inadequate manning levels might be financially adversarial and may even lead to minor and major accidents. An objective and generic approach based on benchmarking results for assessing, evaluating and optimizing operational staffing levels should be standard practice within the chemical industry.

→ *Recommendation 6:*

Operational staffing levels should be looked upon from a "regularly assessment and continuous improvement" viewpoint within plants and multi-plant areas. To this end, a staffing-level-optimizing approach should be employed by plant management and multi-plant management.

Conclusion 7

Top management of different chemical firms within a chemical cluster have to be convinced of the financial benefits of safety and security collaboration.

Every company situated within a chemical cluster faces the risk of being struck by an escalating accident at one of its neighboring plants. These cross-plant risks can be reduced or eliminated if companies are willing to invest in systems and measures to prevent them. The most cost-effective way to invest in external safety and security, is to exchange information between neighboring companies and, based on this information, to calculate the financially most advantageous situation for the cooperating companies.

→ *Recommendation 7:*

A game-theoretic approach to calculate the potential financial benefits of collaboration or non-collaboration (depending on parameters such as external domino-prevention costs, possible plant losses, and accident probabilities) should be employed by top management deciding to strategically cooperate (or not) on safety and security matters with neighboring plants, for example, by taking part (or not) in the multi-plant initiative (suggested and explained in this book) within a chemical multi-plant area.

General Conclusions

To many, a sustainable chemical industry sounds like an oxymoron. How can the industry that stored, produced and handled the substances and materials that caused Mexico City, Bhopal, Seveso, and many other catastrophic accidents ever be sustainable? However, as Rittenhouse (in: Beloff *et al.*, 2005) argues, instead of "*Can* the chemical industry become sustainable?", the question must be "*How* does the chemical industry become sustainable?" An understanding of how companies may take the challenge of building a multi-plant site-integrated safety and security culture and hence contribute to the sustainability of the chemical industry is provided in this book.

The vision that has guided writing this book can be described as the combination of

a) Using all the plant safety and security know-how and knowledge about best practices, case examples, software, *etc.* from the individual plants in a complex industrial area to create multi-plant knowledge bases;

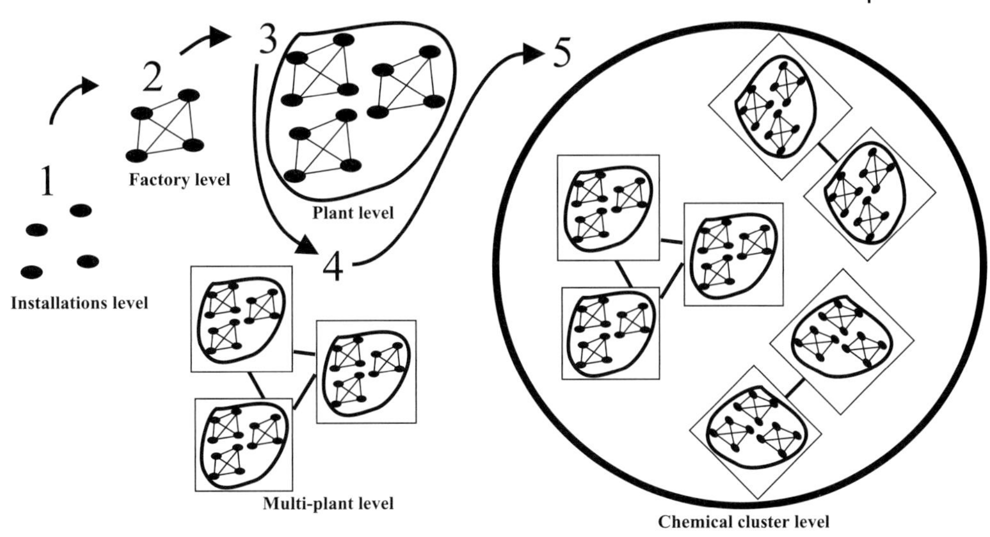

Figure 10.4 The evolution of safety and security management in an industrial area of chemical plants from basic till vision.

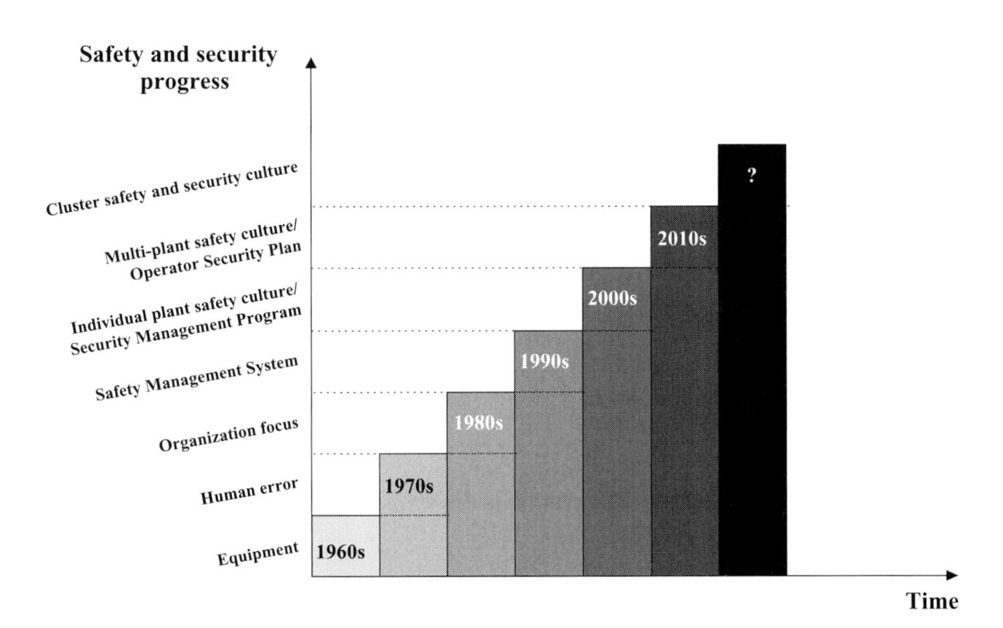

Figure 10.5 The progress of chemical safety and security.

b) Each plant having the ability to create specific versions of these knowledge bases to share detailed safety and security information within the plant about multi-plant safety and security issues and activities.

Figure 10.4 summarizes the safety-management evolution in chemical plants by offering the emerging picture of a cluster characterized by integrated safety.

Figure 10.5 illustrates the continuing progress of safety enhancement in– and of the sustainability of–the chemical industry over time. The figure also depicts security only recently becoming a major issue to be dealt with by chemical plants.

The progression of corporate safety practices–from compliance with government policies to risk management and the developing of a safety-management system, to creating a plant-safety culture, to building integrated safety approaches and systems–can be seen as part of a strategy for ensuring corporate sustainability, finally leading to company leadership. A logical next step is to ensure chemical cluster sustainability by extending corporate social responsibility to an aggregated multi-plant level and including security approaches as well. Lanne and Räikkönnen (in: Brebbia *et al.*, 2005) define safety as freedom from danger, as the condition of being safe, or as the process that limits the safety risk of accident. These authors define security as the extent of protection against some unwanted occurrence such as the corruption of information or deliberate physical damage. Lanne and Räikkönnen further remark that the boundaries of both sectors are not quite clear and that there are a lot of common activities in the sectors. They identify the need for site-integrated safety and security management and illustrate its benefits. Integrated multi-plant safety and security management can indeed be expected in the near future. Figure 10.5 illustrates the latter anticipated evolution. This book closes the gap by making a clear distinction between safety and security and discussing the way to create multi-plant site-integrated safety and security.

There is still much to be done for the vision of an integrated cluster safety and security culture in the chemical industry to achieve its full promise, but the work that has been presented demonstrates that realizing the vision step-by-step is both feasible and desirable. The author hereby invites all stakeholders concerned with ameliorating and optimizing safety and security management of chemical plants and chemical clusters to join the quest to make the vision a reality.

References

Abdolhamidzadeh, B. (2009) Chain of accidents (domino effect) in chemical process industries, working paper.

Acikalin, H.A. (2003) Dynamische Simulation thermisch initiierter Domino-Effecte (in German), PhD thesis, Berlin University, Berlin.

Alberich, R., Miro-Julia, J., and Rossello, F. (2005) *Marvel Universe Looks Almost Like A Real Social Network*, Universitat de les Illes Balears.

Ale, B.J.M. (2005) Living with risk: a management question. *Rel. Eng. Syst. Saf.*, **90**, 196–205.

Alexander, D.C. (2004) *Business Confronts Terrorism: Risks and Responses*, The University of Wisconsin Press, Madison, Wisconsin.

AMINAL (2009) *Manual Failure Frequency Figures for Drafting Safety Reports (MFF), Cel Veiligheidsrapportering (in Dutch)*, Ministry for the Flemish Community.

Antonioni, G., Spadoni, G., and Cozzani, V. (2009) Application of domino effect quantitative risk assessment to an extended industrial area. *J. Loss Prevent. Proc.*, **22** (5), 439–449.

ASIS International (2003) *General Security Risk Assessment Guidelines*, ASIS International, Alexandria, Virginia.

Atkins, W.S. (1998) *Development of Methods to Assess the Significance of Domino Effects from Major Hazard Sites*, Health and Safety Executive, HMSO, London.

Aumann, R.J. (1989) Game theory, in *Game Theory* (eds J. Eatwell, M. Milgate, and P. Newman), The New Palgrave Macmillan, London, pp. 460–482.

Azaiez, M.N., and Bier, V.M. (2007) Optimal resource allocation for security in reliability systems. *Eur. J. Oper. Res.*, **181** (2), 773–786.

Bagster, D.F., and Pitblado, R.M. (1991) The estimation of domino incident frequencies – an approach. *Process Saf. Environ.*, **69** (B), 195–199.

Barabasi, A.L. (2003) *Linked: How Everything Is Connected to Everything Else and What It Means for Business, Science, and Everyday Life*, Plume, New York.

Barabasi, A.L., Albert, R., and Jeong, H. (1999) Mean-field theory for scale-free random networks. *Phys. A*, **272**, 173–187.

Barabasi, A.L., Albert, R., and Jeong, H. (2000) Scale-free characteristics of random networks: the topology of the world-wide web. *Phys. A*, **281**, 69–77.

Baram, M. (1998) Process safety management and the implications of organisational change, in *Safety Management. The Challenge of Change* (eds A. Hale and M. Baram), Pergamon, Oxford, pp. 191–205.

Barrat, A., Barthélemy, M., and Vespignani, A. (2004) Weighted evolving networks: coupling topology and weight dynamics. *Phys. Rev. Lett.*, **92**, 228701.

Barron, E.N. (2008) *Game Theory*, An Introduction, John Wiley & Sons, Inc., Hoboken, New Jersey.

Bellamy, L.J., Geyer, T.A.W., and Wilkinson, J. (2008) Development of a functional model which integrates human factors, safety management systems and wider organizational issues. *Saf. Sci.*, **46**, 461–492.

Multi-Plant Safety and Security Management in the Chemical and Process Industries. G.L.L. Reniers

ISBN: 978-3-527-32551-1

Beloff B., Lines M., and Tanzil D. (eds) (2005) *Transforming Sustainability Strategy into Action. The Chemical Industry*, John Wiley & Sons, Inc., Hoboken, New Jersey.

Bier, V.M., Nagaraj, A., and Abhichandani, V. (2005) Protection of simple series and parallel systems with components of different values. *Reliab. Eng. Syst. Safe.*, **87** (3), 315–323.

Bozzolan, J.C., and Messias de Oliveira Neto, J. (2007) A study on domino effects in nuclear fuel cycle facilities, International Nuclear Atlantic Conference – INAC 2007, Santos, Brazil.

Brebbia, C.A., Bucciarelli, T., Garzia, F., and Guarascio, M. (2005) *Safety and Security Engineering I (SAFE 2005)*, WIT Press, Southampton, UK.

Buchanan, M. (2003) *Nexus; Small Worlds and the Groundbreaking Science of Networks*, W. W. Norton & Company, New York.

Cacciabue, P.C. (2004) *Guide to Applying Human Factors Methods*, Springer-Verlag, London, UK.

Canadian Society for Chemical Engineering (2002) *Process Safety Management*, 3rd edn, Ottawa, Ontario, Canada.

CCPS (Center for Chemical Process Safety) (1992) *Guidelines for Hazard Evaluation Procedures*, 2nd edn, American Institute of Chemical Engineers, New York, USA.

CCPS (Center for Chemical Process Safety) (1993) *Guidelines for Auditing Process Safety Management Systems*, American Institute of Chemical Engineers, New York, USA.

CCPS (Center for Chemical Process Safety) (1994a) *Guidelines for Preventing Human Error in Process Safety*, American Institute of Chemical Engineers, New York, USA.

CCPS (Center for Chemical Process Safety) (1994b) *Guidelines for Implementing Process Safety Management Systems*, American Institute of Chemical Engineers, New York, USA.

CCPS (Center for Chemical Process Safety) (1995) *Tools for Making Acute Risk Decisions*, American Institute of Chemical Engineers (AIChE), New York, USA.

CCPS (Center for Chemical Process Safety) (1996) *Inherently Safer Chemical Processes, A Life Cycle Approach*, American Institute of Chemical Engineers, New York, USA.

CCPS (Center for Chemical Process Safety) (1998) *Local Emergency Planning Committee Guidebook*, American Institute of Chemical Engineers (AIChE), New York, USA.

CCPS (Center for Chemical Process Safety) (2000a) *Evaluating Process Safety in the Chemical Industry: A User's Guide to Quantitative Risk Analysis*, American Institute of Chemical Engineers, New York, USA.

CCPS (Center for Chemical Process Safety) (2000b) *Guidelines for Chemical Process Quantitative Risk Analysis*, 2nd edn, American Institute of Chemical Engineers, New York, USA.

CCPS (Center for Chemical Process Safety) (2001) *Revalidating Process Hazard Analyses*, American Institute for Chemical Engineers, New York, USA.

CCPS (Center for Chemical Process Safety) (2003) *Guidelines for Analyzing and Managing the Security Vulnerabilities of Fixed Chemical Sites*, American Institute of Chemical Engineers, New York, USA.

CCPS (Center for Chemical Process Safety) (2007) *Human Factors Methods for Improving Performance in the Process Industries*, American Institute of Chemical Engineers, New York, USA.

Child, J., Faulkner, D., and Tallman, S. (2005) *Cooperative Strategy: Managing Alliances, Networks and Joint Ventures*, 2nd edn. Oxford University Press, New York, USA.

Churchill, G.A. (1979) A paradigm for developing better measures of marketing constructs. *Journal of Marketing Research*, **16**, 12–27.

Cocchiara, M., Bartolozzi, V., Picciotto, A., and Galluzzo, M. (2001) Integration of interlock system analysis with automated HAZOP analysis. *Rel. Eng. Syst. Saf.*, **74**, 99–105.

Coco, J.C. (2003) *The 100 Largest Losses 1972-2001: Large Property Damage Losses in the Hydrocarbon-Chemical Industries*, Marsh Risk Consulting Practice, New York, USA.

Conrow, E.H. (2003) *Effective Risk Management*, American Institute of Aeronautics and Astronautics (AIAA), Reston, Virginia, USA.

Contractor, F.J., and Lorange, P. (2002) *Cooperative Strategies in International*

Business. Joint Ventures and Technology Partnerships Between Firms, Pergamon, Amsterdam, The Netherlands.

Cooper, M.D. (2000) Towards a model of safety culture. *Saf. Sci.*, **32** (6), 111–136.

Council Directive 96/82/EC (1997) Council Directive 96/82/EC on the control of major-accident hazards involving dangerous substances. *Off. J. Europ. Union*, **L010**, 13–33.

Council Directive 2003/105/EC (2003) Council Directive 2003/105/EC, Seveso II Directive on the control of major-accident hazards involving dangerous substances with amendments. *Off. J. Europ. Union*, **L345**, 97–105.

Covello V.T. (1993) *Merkhofer M.W., Risk Assessment Methods*, Plenum Press, New York, USA.

Cozzani, V., and Salzano, E. (2004) The quantitative assessment of domino effects caused by overpressure Part I: probit models. *J. Hazard. Mater.*, **A107**, 67–80.

Cozzani, V., Gubinelli, G., Antonioni, G., Spadoni, G., and Zanelli, S. (2005) The assessment of risk caused by domino effect in quantitative area risk analysis. *J. Hazard. Mater.*, **A127**, 14–30.

Cozzani, V., Gubinelli, G., and Salzano, E. (2006) Escalation thresholds in the assessment of domino accidental events. *J. Hazard. Mater.*, **129** (1–3), 1–21.

CPR12E Red Book (1997) *Methods for Determining and Processing Probabilities*, 2nd edn, Sdu Uitgevers, The Hague, The Netherlands.

CPR14E Yellow Book (1997) *Methods for the Calculation of Physical Effects*, 3rd edn, Sdu Uitgevers, The Hague, The Netherlands.

CPR16E Green Book (1998) *Methods for Determining the Possible Damage to Humans and Structures Caused by Discharge of Dangerous Goods*, Sdu Uitgevers, The Hague, The Netherlands.

CPR18E Purple Book (1999) *Guidelines for Quantitative Risk Assessment*, Sdu uitgevers, The Hague, The Netherlands.

Crucitti, P., Latora, V., Marchiori, M., and Rapisarda, A. (2003) Efficiency of scale-free networks: error and attack tolerance. *Phys. A*, **320**, 622–642.

Curzio, A.Q., and Fortis, M. (2002) *Complexity and Industrial Clusters. Dynamics and Modcels in Theory and Practice*, Physica-Verlag, Heidelberg, Germany.

Cullen W. (1990) *The Honourable Lord, The Public Inquiry into the Piper Alpha Disaster*, HM Stationary Office, London, UK.

Daft R.L. (2001) *Essentials of Organization Theory & Design*, 2nd edn, South-Western College Publishing, Cincinnati, Ohio, USA.

Delvosalle C. (1996) *Domino Effects Phenomena: Definition, Overview and Classification, European Seminar on Domino Effects, Leuven, Belgium*, Federal Ministry of Employment, Safety Administration, Direction Chemical Risks, Brussels, Belgium.

Delvosalle C. (1998) A methodology for the identification and evaluation of domino effects. Rep. CRC/MT/003, Belgian Ministry of Employment and Labour, Brussels.

De Man, A., Van Der Zee, H., Geurts, D., Kuijt, M., and Vincent, N. (2000) *Competing for Partners*, Pearson Education, Zeist, The Netherlands.

Department of Defence (2000) MIL-STD-882D. Standard Practice for System Safety. HQ Air Force Materiel Command, Ohio.

De Vellis, R.F. (1991) *Scale Development: Theory and Application*, Sage, Newbury Park, CA, US.

Dorman, P. (2000) If Safety Pays, why Don't Employers Invest in it? in *Systematic Occupational Health and Safety Management* (eds K. Frick, P.L. Jensen, M. Quinlan, and T. Wilthagen), Pergamon, Oxford, UK, pp. 351–365.

Early, W.F., and Sherrod, R.M. (1991) Hazard and operability studies, in *Risk Assessment and Risk Management for the Chemical Process Industry* (eds H.R. Greenberg and J.J. Cramer), John Wiley & Sons, Inc, New York, USA, pp. 101–126.

ES1 (2009) MARS (Major Accidents Reporting System) Database, electronic database available via URL http://mahbsrv.jrc.it/mars/Default.html (accessed 05. Feb. 2010).

ES2 (2008) Valuepark website, http://www.dow.com/facilities/europe/benelux/terneuzen/locatie/valuepark.htm (accessed 05. Feb. 2010).

ES3 (2008) Valuepark website, http://www.chemicalparks.com/parks/19/Seiten/default.aspx (accessed 05. Feb. 2010).

ES4 (2008) Kalundborg website, www.symbiosis.dk (accessed 05. Feb. 2010).

ES5: American Chemistry Council (2006) Responsible Care Security Code of Management Practices, available online at www.americanchemistry.com (accessed 05. Feb. 2010).

ES6: American Chemistry Council (2001) Site Security Guidelines for the US Chemical Industry, available online at www.americanchemistry.com (accessed 05. Feb. 2010).

ES7 (2009) Security Guidelines, http://www.asisonline.org/guidelines/guidelines.htm (accessed 05. Feb. 2010).

Fennelly, L.J. (2004) *Handbook of Loss Prevention and Crime Prevention*, 4th edn, Elsevier Butterworth-Heinemann, Oxford, UK.

Fievez, C. (1996) *Effets Domino Dans L'industrie Chimique: Recherche D'une Méthodologie De Prevention Sur Base D'une Analyse Accidentologique*, Faculté Polytechnique de Mons, Mons, Belgium.

Fisher, R.J., Halibozek, E., and Green, G. (2008) *Introduction to Security*, 8th edn, Elsevier Butterworth-Heinemann, Oxford, UK.

Ford, K.A., and Brown, W.H. (1990) Innovative applications of the Hazop Technique. Presentation at the AIChE Spring National Meeting, Orlando, Florida, USA.

Frick, K., Jensen, P.L., Quinlan, M., and Wilthagen, T. (2000) *Systematic Occupational Health and Safety Management*, Pergamon, Oxford, UK.

Gardner, R.J., and Reyne, M.R. (1994) *Selection of Safety Interlock Integrity Levels*, Dupont Engineering.

Gerling, D.W., and Anderson, J.C. (1988) An updated paradigm for scale development incorporating unidimensionality and its assessment. *Journal of Marketing Research*, **25**, 186–192.

Golany, B., Kaplan, E.H., Marmur, A., and Rothblum, U.G. (2009) Nature plays with dice – terrorists do not: allocating resources to counter strategic versus probabilistic risks. *Eur. J. Oper. Res.*, **192** (1), 198–208.

Gong, P., and van Leeuwen, C. (2003) Emergence of scale-free network with chaotic units. *Phys. A*, **321**, 679–688.

Goossens, L.H.J. (2004) Risk and vulnerability of critical infrastructures. *J. Risk Res.*, **7** (6), 567–568.

Gorrens, B., De Clerck, W., De Jongh, K., and Aerts, M. (2009) Domino effecten van en naar Seveso-inrichtingen. Rep. 07.0007, Flemish Ministry of Environment, Nature and Energy, Brussels.

Greenberg, H.R., and Cramer, J.J. (1991) *Risk Assessment and Risk Management for the Chemical Process Industry*, John Wiley & Sons, Inc., New York, USA.

Gruhn, P. (2004) *Safety Instrumented System Design: Lessons Learned. Process Safety Progress*, **18** (3), 156–160.

Gubinelli, G., Zanelli, S., and Cozzani, V. (2004) A simplified model for the assessment of the impact probability of fragments. *J. Hazard. Mater.*, **A116**, 175–187.

Hale, A.R., Wilpert, B., and Freitag, M. (1997) *After the Event, from Accident to Organisational Learning*, Pergamon, Oxford, UK.

Hausken, K. (2008) Strategic defense and attack for reliability systems. *Reliab. Eng. Syst. Safety*, **93** (11), 1740–1750.

Hausken, K., and Levitin, G. (2009) Minimax defense strategy for complex multi-state systems. *Reliab. Eng. Syst. Safety*, **94** (2), 577–587.

HSE (Health and Safety Executive) (1984) *The Control of Major Hazards, Third Report of the HSC Advisory Committee on Major Hazards*, HMSO, London, UK.

HSE (Health and Safety Executive) (2004) *Safe Staffing Arrangements*, HSE Publications, London, UK.

HSE (Health and Safety Executive) (2005a) *Human Factors in the Management of Major Accident Hazards*, HSE Publications, London, UK.

HSE (2005b) Research Report 367, A review of safety culture and safety climate literature for the development of the safety culture inspection toolkit. HMSO, London, UK.

Heal, G., and Kunreuther, H. (2007) Modeling interdependent risks. *Risk Anal.*, **27** (3), 621–634.

Hemre, A. (2005) Building and sustaining communities of practice at ericsson research Canada, in *Knowledge Management Tools and Techniques. Practitioners and Experts Evaluate KM Solutions* (ed. Rao, M.), Butterworth-Heinemann, Oxford, UK, pp. 155–165.

Hirst, I.L., and Carter, D.A. (2002) A "worst case" methodology for obtaining a rough but rapid indication of the societal risk from a major accident hazard installation. *J. Hazard. Mater.*, **92**, 223–237.

Hopkins, A., and Hale, A. (2002) Issues in the regulation of safety: setting the scene, in *Changing Regulation* (eds. M. Kirwan, A. Hale and A. Hopkins), Pergamon, Oxford, UK, The Netherlands, pp. 1–12.

Hovden, J. (1998) Models of Organizations versus Safety Management Approaches: a discussion based on studies of the "internal control of SHE" reform in Norway, in *Safety Management. The Challenge of Change* (eds A. Hale and M. Baram), Pergamon, Oxford, UK, pp. 23–41.

Howard, N. (1966) The theory of metagames, General Systems 11(V).

IAEA, International Atomic Energy Agency (1998) INFCIRC/225/Rev. 4.

INERIS (2002) *Institut National De L'environment Industrial Et Des Risques, Méthode Pour L'identification Et La Caractérisation Des Effets Dominos MICADO® (in French)*, Direction des Risques Accidentels, Paris, France.

International Standard (2003) IEC-61511, 1st edn. International Electrotechnical Commission.

Ioannidis, C., Logothetis, C., Potsiou, C., Kiranoudis, C., Christolis, M., and Markatos, N. (1999) Spatial information management for risk assessment of major industrial accidents. Proceedings (CD) of F.I.G. Commission 3 Annual Meeting and Seminar on 'Spatial Information Management', Budapest, Hungary.

ISO Guide 72 (2001).*Guidelines for the Justification and Development of Management System Standards*, ISO.

Kahneman, D., and Tversky, A. (1984) Choices, values and frames. *Am. Psychol.*, **39**, 341–350.

Kelly, A. (2004) *Decision Making Using Game Theory*, Cambridge University Press, Cambridge, UK.

Khan, F.L., and Abbasi, S.A. (1998a) DOMIFFECT (DOMIno eFFECT): a new software for domino effect analysis in chemical process industries. *Environ. Model. Software*, **13**, 163–177.

Khan, F.I., and Abbasi, S.A. (1998b) Models for domino effect analysis in chemical process industries. *Process Saf. Prog.*, **17** (2), 107–113.

Khan, F.I., and Abbasi, S.A. (1999) Assessment of risks posed by chemical industries – application of a new computer automated tool MAXCRED-III. *J. Loss Prevent. Proc.*, **12** (6), 455–469.

Khan, F.I., and Abbasi, S.A. (2001) An assessment of the likelihood of occurrence, and the damage potential of domino effect (chain of accidents) in a typical cluster of industries. *J. Loss Prevent. Proc.*, **14**, 283–306.

Kirchsteiger, C. (1998) Absolute and relative ranking approaches for comparing and communicating industrial accidents. *J. Hazard. Mater.*, **59** (1), 31–54.

Kirchsteiger, Ch. (2003) Industrial risk management and international agreements. *Int. J. Emerg. Man.*, **1** (3), 247–267.

Kirwan, B., Hale, A., and Hopkins, A. (2002) *Changing Regulation, Controlling Risks in Society*, Pergamon, Oxford, UK.

Kletz, T.A. (1998) in *What Went Wrong: Case-Studies of Process Plant Disasters*, 4th edn, Butterworth-Heinemann, Oxford, UK.

Kletz, T.A. (1999) *Hazop and Hazan*, Editions Technip, Paris.

Landoll, D.J. (2006) *The Security Risk Assessment Handbook*, Auerbach Publications (Taylor & Francis Group), Boca Raton, USA.

Latora, V., and Marchiori, M. (2001) Efficient behavior of small-world networks. *Phys. Rev. Lett.*, **87** (19).

Latora, V., and Marchiori, M. (2002) Is the Boston subway a small-world network? *Phys. A*, **314**, 109–113.

Latora, V., and Marchiori, M. (2003) Economic small-world behaviour in weighted networks. *Eur. Phys. J. B*, **32**, 249–263.

Lees, F.P. (1996) *Loss Prevention in the Process Industries*, 2nd edn, Butterworth-Heinemann, Oxford, UK.

Liu, D., Wang, X., and Camp, J. (2008) Game-theoretic modeling and analysis of insider threats. *Int. J. Crit. Infrastr. Prot.*, **1**, 75–80.

Luu, V.T., Kim, S.Y., and Huynh, T.A. (2008) Improving project management performance of large contractors using benchmarking approach. *Int. J. Proj. Management*, **26** (7), 758–769.

Mahoney, D.G. (1990) in *Large Property Damage Losses in the Hydrocarbon-Chemical Industries: A Thirty-Year Review*, 13th edn, M&M Protection Consultants, New York, USA.

Malone, T.W., Crowston, K., and Herman, G.A. (2003) *Organizing Business Knowledge. The MIT Process Handbook*, The MIT Press, Cambridge, Massachusetts, USA.

Marszal, E.M., Fuller, B.A., and Shah, J.N. (1999) *Comparison of Safety Level Selection Methods and Utilization of Risk Based Approaches*, Four Elements Inc., Columbus, Ohio, USA.

Milgrom, P., and Roberts, J. (1990) Rationalizability, learning and equilibrium in games with strategic complementarities. *Econometrica*, **58** (6), 1255–1277.

MI5, Security Service UK (2005) *Protection Against Terrorism*, Crown, London, UK.

Mol, T. (2003) *Productive Safety Management*, Butterworth-Heinemann, Oxford, UK.

Montet, C., and Serra, D. (2003) *Game Theory and Economics*, Palgrave MacMillan, New York.

Moore, D.A. (2004) The new risk paradigm for chemical process security and safety. *J. Hazard. Mater.*, **115**, 175–180.

Moore, D.A., Fuller, B., Hazzan, M., and Jones, J.W. (2005) U.S. Department of Homeland Security; Risk Analysis Methodology for Critical Asset Protection (RAMCAP) Project. International Seminar RISK: Perception, Communication, Acceptability, Bruges, Belgium, October 3–4, 2005.

Moore, D.A., Fuller, B., Hazzan, M., and Jones, J.W. (2007) Development of a security vulnerability assessment process for the RAMCAP chemical sector. *J. Hazard. Mater.*, **142** (3), 689–694.

Murnighan, J.K., and Mowen, J.C. (2002) *The Art of High-Stakes Decision-Making*, John Wiley & Sons, Inc., New York, USA.

National Research Council (1994) *Committee on Risk Assessment of Hazardous Air Polluants, Science and Judgment in Risk Assessment*, National Academy Press, Washington, DC.

OECD (2003) *Guiding Principles for Chemical Accident Prevention, Preparedness and Response*, 2nd edn, Organization for Economic Cooperation and Development, Paris, France.

OECD (2004) *OECD General Guidance on Risk Management Programs for Chemical Accidents Prevention*, Organization for Economic Cooperation and Development, Paris, France.

Oh, J.I.H. (2002) The EU seveso II Directive: an example of a regulation that could act as an initiator to raise the major hazard safety awareness within industry, in *Changing Regulation* (eds. B. Kirwan, A. Hale and A. Hopkins), Pergamon, Oxford, UK, pp. 43–54.

OPITSC (2001) *Code of Practice on Safety Management Systems for the Chemical Industry*, Oil and Petrochemical Industry Technical and Safety Committee, & Occupational Safety Department, Ministry of Manpower, Singapore.

Papazoglou, I.A., Bellamy, L.J., Hale, A.R., Aneziris, O.N., Ale, B.J.M., Post, J.G., and Oh, J.I.H. (2003) I-Risk: development of an integrated technical and management risk methodology for chemical installations. *J. Loss Prevent. Proc.*, **16**, 575–591.

Parmenter, D. Key Performance Indicators. (2007). *Developing, Implementing and Using Winning Kpis*, John Wiley & Sons, Inc., Hoboken, New Jersey, USA.

Pettitt, G.N., Schumacher, R.R., and Seeley, L.A. (1993) Evaluating the probability of major hazardous incidents as a result of escalation events. *J. Loss Prevent. Proc.*, **6** (1), 37–46.

Pidgeon, N., and O'Leary, M. (2000) Man-made disasters: why technology and organizations (sometimes) fail. *Saf. Sci.*, **34** (1–3), 15–30.

Pietersen, C.M. (1986) Analysis of the LPG disaster in Mexico City. *Loss Prev. Saf. Prom.*, **5**, 21.

Pietersen, C.M. (1990) Consequence of Accidental Release of Hazardous Materials. *J. Loss Prevent. Proc.*, **3**, 136–141.

Post, J.G., Bottellerghs, P.H., Vÿgen, L.J., and Matthÿsen, A.J.C.M. (2003) Instrument Domino Effecten, RiVM, Bilthoven, The Netherlands.

Powell, R.L. (1996) Process Safety and Control Systems Integrity. International Conference and Workshop on Process Safety Management and Inherently Safer Processes, October 8-11, 1996, Orlando, Florida, New York, American Institute for Chemical Engineers, USA, pp. 227–241.

Preston, M.L., and Swann, C.D. (1995) Twenty-five years of hazops. *J. Loss Prevent. Proc.*, **8** (6), 349–353.

Raghavan, K.V., and Swaminathan, G. (1996) *Hazard Assessment and Disaster Mitigation in Petrochemical Industries, Chennai,* Oxford Publishing Company, UK.

Rampersad, H. (2004) *Total Performance Scorecard*, 2nd edn, Scriptum, Schiedalm, The Netherlands.

Rao, M. (2005) *Knowledge Management Tools and Techniques. Practitioners and Experts Evaluate KM Solutions*, Butterworth-Heinemann, Oxford, UK.

Rapoport, A., and Guyer, M. (1966) A taxonomy of 2x2 games. *Gen. Syst.*, **11** (5), 203–214.

Reason, J.T. (1990) *Human Error*, Cambridge University Press, Cambridge, UK.

Reason, J. (1997) *Managing the Risks of Organizational Accidents*, Ashgate Publishing Limited, Aldershot, UK.

Reniers, G., and Soudan, K. (2003) Risicoanalyse procedures in de scheikundige nijverheid: resultaten van kwalitatief onderzoek bij 24 chemische plants, Economisch en Sociaal Tijdschrift (voorloper Kwartaalschrift Economie). *Universiteit Antwerpen*, **57** (3), 249–274.

Reniers, G.L.L., Dullaert, W., Ale, B.J.M., and Soudan, K. (2005a) The use of current risk analysis tools evaluated towards preventing exterFnal domino accidents. *J. Loss Prev. Process Ind.*, **18** (3), 119–126.

Reniers, G.L.L., Dullaert, W., Ale, B.J.M., and Soudan, K. (2005b) Developing an external domino accident prevention framework: hazwim. *J. Loss Prev. Process Ind.*, **18** (3), 127–138.

Reniers, G., Ale, B.J.M., Dullaert, W., and Foubert, B. (2006) Decision support systems for major accident prevention in the chemical process industry: a developers' survey. *J. Loss Prev. Process Ind.*, **19** (6), 604–620.

Reniers, G., Dullaert, W., Ale, B., Verschueren, F., and Soudan, K. (2007a) Engineering an instrument to evaluate safety critical manning arrangements in chemical industrial areas. *J. Bus. Chem.*, **4** (2), 60–75.

Reniers, G.L.L., and Dullaert, W. (2007b) DomPrevPlanning: user-friendly software for planning domino effects prevention. *Saf. Sci.*, **45**, 1060–1081.

Reniers, G.L.L., Dullaert, W., Audenaert, A., Ale, B.J.M., and Soudan, K. (2008a) Managing domino effect-related security of industrial areas. *J. Loss Prevent. Proc.*, **21**, 336–343.

Reniers, G.L.L., and Dullaert, W. (2008b) Knock-on accident prevention in a chemical cluster. *Expert Syst. Appl.*, **34**, 42–49.

Reniers, G.L.L., Ale, B.J.M., Dullaert, W., and Soudan, K. (2009a) Designing continuous safety improvement within chemical industrial areas. *Saf. Sci.*, **47** (5), 578–590.

Reniers, G., Dullaert, W., and Soudan, K. (2009b) Shaping a novel security approach in chemical industrial clusters to prevent large-scale domino events. *Secur. J.*, **22** (2), 119–142.

Reniers, G., Dullaert, W., and Soudan, K. (2009c) Domino effects within a chemical cluster: a game-theoretical modeling approach by using Nash-equilibrium. *J. Hazard. Mater.*, **167** (1–3), 289–293.

Reniers, G.L.L., and Audenaert, A. (2009d) Chemical plant innovative safety investments decision-support methodology. *Journal of Safety Research*, **40** (6), 411–419.

Reniers, G. (2009e) How to increase multi-plant collaboration within a chemical cluster and its impact on external domino effect cooperation initiatives in *Safety and Security Engineering III*. (eds. M. Guarascio, C.A. Brebbia, F. Garzia), WIT Press, pp. 379–388.

Reniers, G. (2009f) Terrorism security in the chemical industry: results of a qualitative investigation. *Secur. J.*, **48** (2), 179–185. DOI: 10.1057/SJ.2009.10.

Reniers, G.L.L. (2010a) A novel methodology for evaluating the change of existing operational staffing levels (MCSL) in the chemical industry. *Saf. Sci.*, forthcoming.

Reniers, G., and Soudan, K. (2010b) A game-theoretical approach for reciprocal security-related prevention investment decisions. *Rel. Eng. Syst. Saf.*, **95** (1), 1–9.

Reniers, G.L.L. (2010c) Optimizing and balancing operational manning levels and SHEQ within chemical companies. *J. Loss Prevent. Proc.*, **23**, 60–70.

Ridley, J., and Channing, J. (2003) *Safety at Work*, 6th edn, Butterworth-Heinemann, Oxford, UK.

RIVM (2003) *Rijksinstituut Voor Volksgezondheid En Milieu*, Instrument Domino Effecten, Bilthoven, The Netherlands.

Salzano, E., and Cozzani, V. (2005) The analysis of domino accidents triggered by vapor cloud explosions. *Rel. Eng. Syst. Saf.*, **90**, 271–284.

Schein, E. (1990) Organizational culture. *Am. Psychol.*, **45**, 109–119.

Schreiber, G., Akkermans, H., Anjewierden, A., de Hoog, R., Shadbolt, N., Van de Velde, W., and Wielinga, B. (2001) *Knowledge Engineering and Management*, 2nd edn, MIT Press, Cambridge, Massachusetts, USA.

Sinnott, R.K. (1996) Safety and loss prevention, in *Coulson & Richardson's Chemical Engineering* vol. **6**, 2nd edn (ed. R.K. Sinnott), Butterworth Heinemann, London, UK, pp. 319–356.

Smith, J. (1988) *Commun. ACM*, **31**, 1202.

Smith, R. (2007) *Business Process Management and the Balanced Scorecard. Using Processes as Strategic Drivers*, John Wiley & Sons, Inc., Hoboken, New Jersey, USA.

Smith, D.J., and Simpson, K.G.L. (2005) *Functional Safety: A Straightforward Guide to Applying IEC 61508 and Related Standards*, 2nd edn, Butterworth-Heinemann, Oxford, UK.

Spadoni, G., Egidi, D., and Contini, S. (2000) Through ARiPAR-GiS the quantified area risk analysis supports land–use planning activities. *J. Hazard. Mater.*, **71** (1–3), 423–437.

Spadoni, G., Contini, S., and Uguceioni, G. (2003) The new version of ARiPAR and the benefits given in assessing and managing major risks in industrialized areas. *Process Safety and Environmental Protection*, **81** (1), 19–30.

Sprague, R.H. Jr., and Watson, H.J. (1986) *Decision Support Systems: Putting Theory into Practice*, Prentice Hall, London, UK.

Standards Australia/Standards New Zealand (1999) AS/NZS4360. Risk Management, Australian/New Zealand Standard.

Stephans, R.A. (2004) *System Safety for the 21st Century*, John Wiley & Sons, Inc., Hoboken, New Jersey, USA.

Sutherland, V., Makin, P., and Cox, C. (2000) *The Management of Safety*, SAGE Publications Ltd, London, UK.

Swafford, P.M., Ghosh, S., and Murthy, N. (2006) The antecedents of supply chain agility of a firm: scale development and model testing. *Journal of Operation Management*, **24** (2), 170–188.

Thurston, C.W. (1994) Automation in chemical plant safety: a design philosophy, in *International Symposium and Workshop on Safe Chemical Process Automation*, Center for Chemical Process Safety (CCPS), Health and Safety Executive (HSE), Chemical Manufacturers Association (CMA), IEC, ISA, Houston, Texas. September 27-29, pp. 4–29.

Tsoukas, H., and Shepherd, J. (2004) *Managing the Future: Foresight in the Knowledge Economy*, Blackwell Publishing, Oxford, UK.

Turicchi, S., Rosen, B., Grossel, S., and Safety (2000) *Quality and Continuous Improvement–Applying the Quality Process in Chemical Process Safety Management*, Joint EFCOG/DOE Chemical safety Management Workshop, Washington DC.

Turner, B.A., and Pidgeon, N.F. (1997) *Man-Made Disasters*, 2nd edn, Butterworth-Heinemann, London, UK.

Tweeddale, M. (2003) *Managing Risk and Reliability of Process Plants, Gulf Professional Publishing*, Butterworth-Heinemann, Burlington, Massachusetts, USA.

United States Government Accountability Office, Dep. Homeland Security (2006) *DHS Is Taking Steps to Enhance Security at Chemical Facilities, but Additional Authority Is Needed*, GAO-06-150, United States Government Accountability Office, Dep. Homeland Security.

Vallee, A., Bernuchon, E., and Hourtolou, D. (2002) MICADO: Méthode pour l'identification et la caractérisation des effets dominos. Rep. INERIS-DRA-2002-25472, Direction des Risques Accidentels, Paris, France.

Vega-Redondo, F. (2003) *Economics and the Theory of Games*, Cambridge University Press, Cambridge, UK.

Watts, D.J. (2004) *Small Worlds*, Princeton University Press, New Jersey, USA.

Wells, G. (1997) *Major Hazards and Their Management*, Institution of Chemical Engineers, Rugby, UK.

Wettig, J., Porter, S., and Kirchsteiger, C. (1999) Major industrial accidents regulation in the European Union. *J. Loss Prev. Proc.*, **12** (1), 19–28.

Williamson, A.M., Feyer, A.M., Cairns, D., and Biancotti, D. (1997) The development of a measure of safety climate: the roles of safety perceptions and attitudes. *Saf. Sci.*, **25**, 15–27.

Wilpert, B., and Fahlbruch, B. (2002) *System Safety, Challenges and Pitfalls of Intervention*, Pergamon, Oxford, UK.

Zwetsloot, G.I.J.M., Gort, J., Steijger, N., and Moonen, C. (2007) Management of change: lessons learned from staff reductions in the chemical process industry. *Saf. Sci.*, **45** (7), 769–789.

Appendix A Instrument for Evaluating Security Staffing Levels

1. Supervision / Intervention possibility (Detection)	Yes	No
1.1. Is continuous supervision of all critical infrastructure possible (24/7)		
A. by outdoor security guards (outdoor surveillance and interventions)?...	☐	☐
B. by indoor security guards (indoor surveillance)?	☐	☐
C. by a combination of indoor and outdoor security guards?	☐	☐

C.Q.: In which way can this be guaranteed (qualitatively as well as quantitatively)?

e.g.

- a team of all-round competent outdoor security staff is assigned to be permanently present at all critical infrastructure in the industrial area
- a team of outdoor security staff is assigned to guard a part of a critical infrastructure and a part of critical infrastructure is permanently under surveillance by indoor security guards
- a team of indoor security staff is responsible for surveillance of all critical infrastructure which is permanently monitored by surveillance equipment

If one of the answers is rated "yes", ranking 1. =1A, go to 2.1; otherwise go to 1.2.

Multi-Plant Safety and Security Management in the Chemical and Process Industries. G.L.L. Reniers

ISBN: 978-3-527-32551-1

1.2. Is supervision of all critical infrastructure possible on a semi-continuous basis (e.g., outdoor security guards have a rotating surveillance system) with the condition that all these infrastructures are under surveillance at least 21 hours a day, 7 days a week

A. by outdoor security guards (outdoor surveillance and interventions)?...	☐	☐
B. by indoor security guards (indoor surveillance)?	☐	☐
C. by a combination of indoor and outdoor security guards?	☐	☐

C.Q.: Does the possibility exist that the 21 hours per day surveillance may not be achieved?

If one of the answers is rated "yes", ranking 1. =1B, go to 2.1; otherwise go to 1.3.

1.3. Is supervision of all critical infrastructure possible on a semi-continuous basis with the condition that all these infrastructures are under surveillance at least 18 hours a day, 7 days a week

A. by outdoor security guards (outdoor surveillance and interventions)?...	☐	☐
B. by indoor security guards (indoor surveillance)?	☐	☐
C. by a combination of indoor and outdoor security guards?	☐	☐

C.Q.: Does the possibility exist that the 18 hours per day surveillance may not be achieved?

If one of the answers is rated "yes": ranking 1. =1C; otherwise ranking 1. =1D; go to 2.1.

Additional documents to consult in order to evaluate principle 1: security shift team system, work schemes, technical information concerning the surveillance equipment, *etc.*

2. Distractions (Detection)	**Yes**	**No**

2.1. Do the security guards have to perform / attend to other specific tasks (e.g., answering phone, administration, production tasks, *etc.*) in addition to surveillance tasks

A. outdoor (outdoor surveillance)?	☐	☐
B. indoor (indoor surveillance)?	☐	☐

C.Q.: If the answer is "no", how can this be guaranteed?

If both answers are rated "no", ranking 2. =2A, go to 3.1; otherwise go to 2.2.

2.2. Is there provision for an alternative should the security guards miss the identification of a potential threatening situation due to other tasks

A. outdoor (outdoor surveillance)?	☐	☐
B. indoor (indoor surveillance)?	☐	☐

C.Q.: What is the alternative? (e.g., there are other security guards on duty, installation operators are present at the critical assets location, *etc.*)

If both answers are rated "yes", go to 2.3; otherwise ranking 2. =2D, go to 3.1.

2.3. Is the time left to respond to the threatening situation sufficient?

		Yes	No
A.	outdoor (outdoor surveillance)?	☐	☐
B.	indoor (indoor surveillance)?	☐	☐

C.Q.: On which basis is the time margin defined in the industrial area?

If both answers are rated "yes": ranking 2. =2B if time margin < 5 min; ranking 2. =2C if time margin exceeds 5 min; otherwise ranking 2. =2D; go to 3.1.

Additional documents to consult in order to evaluate principle 2: listing of (other) security guard tasks, technical information concerning the alternative alarm, technical information concerning the time definitions within the company, *etc.*

3. Information (Diagnosis)	**Yes**	**No**

3.1. Have training sessions (including knowledge about hazardous substances and know-how on the prevention of intentional undesirable acts where chemical substances are involved) about how to react in various threatening situations been given

		Yes	No
A.	to all outdoor security guards (outdoor surveillance and interventions)?	☐	☐
B.	to all indoor security guards (indoor surveillance)?	☐	☐

C.Q.: If both answers are 'yes', what is the guarantee that training sessions (a) cover all possible threatening situations and (b) meet high enough technical and qualitative standards?

Go to 3.2.

3.2. Are training sessions periodically repeated and regularly attuned to the latest insights in the security field

		Yes	No
A.	in the case of outdoor security guards (outdoor surveillance and interventions)?	☐	☐
B.	in the case of indoor security guards (indoor surveillance)?	☐	☐

C.Q.: What is the guarantee that training is up-to-date?

If both questions 3.1 and 3.2 are rated "yes" for all topics, ranking 3. =3A, go to 4.1; otherwise go to 3.3.

3.3. Have training sessions about how to react in various threatening situations been given

		Yes	No
A.	to a limited number of outdoor security guards (outdoor surveillance and interventions)?	☐	☐
B.	to a limited number of indoor security guards (indoor surveillance)?	☐	☐

C.Q.: If both answers are "yes", what is the guarantee that training sessions (a) cover all possible threatening situations and (b) have high enough technical and qualitative standards?

Go to 3.4.

3.4. Are trained guards present together with non-trained guards at all times (24/7)

	Yes	No
A. outdoor (physical interventions)?	☐	☐
B. indoor (computer-related interventions)?	☐	☐

C.Q.: Explain how the teams are formed and how continuous presence can be guaranteed.

If both questions 3.3 and 3.4 are rated "yes" for all topics, ranking 3. =3B, go to 4.1; otherwise go to 3.5.

3.5. Is there an alternative for non-trained guards to be informed within minutes

	Yes	No
A. outdoor (physical interventions)?	☐	☐
B. indoor (computer-related interventions)?	☐	☐

C.Q.: Explain the alternative solution. (e.g., guard know-how within cluster center continuously available.

If both answers are rated "yes", ranking 3. =3C; otherwise ranking 3. =3D, go to 4.1.

Additional documents to consult in order to evaluate principle 3: listing of team arrangements, technique of attainability, technical training information, technical information concerning the alternative solution, *etc.*

4. Communication links (Diagnosis)	**Yes**	**No**

4.1. To diagnose and to solve certain security problems, is a specific kind of communication link required

	Yes	No
A. between outdoor security guards?	☐	☐
B. between indoor security guards?	☐	☐
C. between indoor security guards and outdoor security guards?	☐	☐

C.Q.: How can this be guaranteed?

If all three answers are rated "no", ranking 4. =4A, go to 5.1; otherwise go to 4.2.

4.2. Can this communication link be guaranteed 24/7?

	Yes	No
A. between outdoor security guards?	☐	☐
B. between indoor security guards?	☐	☐
C. between indoor security guards and outdoor security guards?	☐	☐

C.Q.: How can it be guaranteed?

If all three answers are rated "yes", ranking 4. =4B, go to 5.1; otherwise go to 4.3.

4.3. Does a back-up system exist consisting of one or more communication links within the plant/cluster that can be guaranteed 24/7 for usage by security guards in abnormal situations (NOT necessarily emergency situations!)?

C.Q.: Explain the back-up system.

If the answer is rated "yes", ranking 4. =4C, otherwise ranking 4. =4D, go to 5.1.

Additional documents to consult in order to evaluate principle 4: listing of communication links, technical specifications concerning the communication links and their reliability, technical information concerning the back-up system, *etc.*

5. Assisting personnel (Handling)	**Yes**	**No**
5.1. To handle certain security problems, is manning assistance needed		
A. outdoor (physical interventions)?	☐	☐
B. indoor (computer-related interventions)?	☐	☐

C.Q.: If the answer is 'no', how can this be guaranteed?

If both answers are rated "no", ranking 5. =5A, go to 6.1; otherwise go to 5.2.

5.2. Is this security manning assistance available 24/7		
A. outdoor (physical interventions)?	☐	☐
B. indoor (computer-related interventions)?	☐	☐

C.Q.: How can this be guaranteed?

Go to 5.3.

5.3. Can this security manning assistance arrive in time		
A. outdoor (physical interventions)?	☐	☐
B. indoor (computer-related interventions)?	☐	☐

C.Q.: How is this time margin defined? Explain.

If both questions 5.2 and 5.3 are rated "yes" for all topics, go to 5.4; otherwise ranking 5. =5D, go to 6.1.

5.4. Does an alternative exist if the manning assistance fails to arrive in time

A. outdoor (physical interventions)? .... ☐ ☐

B. indoor (computer-related interventions)? .... ☐ ☐

C.Q.: Explain this alternative.

If the answers to questions 5.2, 5.3 and 5.4 are all rated "yes": ranking 5. =5B if time in which availability is ensured is less than 5 min; ranking 5. =5C if time in which availability is ensured exceeds 5 min; otherwise ranking 5. =5D; go to 6.1.

Additional documents to consult in order to evaluate principle 5: shift team system, manning assistance system, technical information about attainability of manning assistance, time-margin calculations, alternative information, *etc.*

---

6. Handling operations (Handling)	**Yes**	**No**

6.1. Can handling the threatening situation be accomplished within the minimum calculated time margin

A. outdoor (physical interventions)? .... ☐ ☐

B. indoor (computer-related interventions)? .... ☐ ☐

C.Q.: How can this be guaranteed? (e.g., task analyses, desktop exercices, simulations, *etc.*)

If both answers are rated "yes", go to 6.2.; otherwise go to 6.5.

6.2. Is it possible that the security guards

A. are assigned extra problems to handle (e.g.: back-up activities of other security guards within the plant/cluster)? .... ☐ ☐

B. are assigned extra tasks (e.g.: responsibility for site alarm, emergency phone services, responsibility for low-threat situations)? .... ☐ ☐

C.Q.: If the answer is "no", how can this be guaranteed?

If both answers are rated "no", ranking 6. =6A, stop audit; otherwise go to 6.3.

6.3. Are they timely informed about the extra assignments?

C.Q.: How can this be guaranteed?

If the answer is rated "yes", go to 6.4; otherwise ranking 6. =6D, stop evaluation.

6.4. Can all assigned tasks be accomplished (originally assigned handling situations, extra assigned handling situations and extra tasks)

A. by outdoor security guards (outdoor surveillance)? .... ☐ ☐

B. by indoor security guards (indoor surveillance)? .... ☐ ☐

C.Q.: How can this be guaranteed?

If both answers are rated "yes", ranking 6. =6B, stop audit; otherwise go to 6.5.

6.5. Does a back-up solution (at plant/cluster level) exist should it prove impossible to accomplish the tasks

A. for outdoor security guards (outdoor surveillance)? .... ☐ ☐

B. for indoor security guards (indoor surveillance)? .... ☐ ☐

C.Q.: Explain the back-up system. (e.g., extra back-up guards, assistance personnel to handle the administrative tasks and technical support)

If both answers are rated "yes", ranking 6. =6C; otherwise ranking 6. =6D; stop evaluation.

Additional documents to consult in order to evaluate principle 6: handling problem time-margin calculations/simulations, technical recovery information/data, back-up system data, *etc.*

Appendix B The IESLA Instrument

B.1 Shift System, Flexibility and Organization

In the case of continuous operations within the organization, a sound shift system where flexibility exists within the shifts and between the shifts is essential for optimized manning levels. Furthermore, the manning levels need to take unexpected situations, circumstances and events into account. It would be optimal for operational teams to be organized in such a way that functions within the teams can be filled in by either functions of other teams within the organization (e.g., from another business unit), or they can be filled in by other function(s) within the team. Hence, questions for this category are characterized by manning forming a flexible and reliable entity within an organization or within an organizational unit.

Shift System, Flexibility and Organization: Ideal Manning

- The organizational shift system is able to carry out all (different types of) tasks under various circumstances.
- All responsibilities, functions and tasks within the organizational shift system are unambiguously known by all personnel concerned and this for various circumstances.

Shift System, Flexibility and Organization: Measured Manning

- **Flexibility (/adaptability):** The different shifts and the way they are composed (functions, qualities, etc.) are designed in such manner that it is possible to carry out all required tasks in various circumstances.
- **Logic and consistency:** The way the responsibilities are established within the organization and the way management and production tasks are organized in various circumstances, makes sense.

Questions:

1. How does the organization ensure that the functions and tasks within a shift, if necessary, can be completed in various situations / circumstances?

Multi-Plant Safety and Security Management in the Chemical and Process Industries. G.L.L. Reniers

ISBN: 978-3-527-32551-1

Measures:

- *For each function within a shift, if required, i.e. in case the person being responsible for the function is unavailable for some reason or another, there is a possibility to replace this person.*
- *Other employees with the same qualifications as those of the unavailable person can always be contacted, and if required, can reach the company (relatively) quickly.*

2. How does the organization ensure that a shift has all the necessary competencies to be able to complete the tasks under various circumstances?

   Measures:

   - *For each shift team a competencies' scheme is drafted.*
   - *The competencies' scheme is maintained and evaluated shift-by-shift by installation, business unit or company management in order to guarantee adequate task performance in various situations and circumstances.*

3. How does the organization ensure the filling in of functions (regarding function profiles, tasks, *etc.*) to remain up-to-date?

   Measures:

   - *For each function within every shift a function profile is introduced, which, if required (e.g., in the case of changes), could be adjusted by company, business unit or installation management. Every person carrying out a certain function is aware of his/her function profile.*
   - *For each function within every shift a minimum package of tasks is drafted, which, if necessary (e.g., in the case of changes), could be adjusted by company, business unit or installation management. Every person carrying out a certain function is aware of his/her minimum package of tasks.*

B.2 Technology and Task Complexity

Manning levels strongly depend on the technology available (or not) within the organization to help carrying out the tasks. 'Technology' refers to displays, alarms, control screens, computers, manual equipment, *etc.* Interaction and use of technology can be taken into account by the methodology by verifying the conformity between manning features, on the one hand, and task and technology characteristics, on the other hand.

Technology and Task Complexity: Ideal Manning

- The organization's technology can be employed (by the available manning levels) in various circumstances.
- The tasks' complexities form by no means an obstacle for carrying out the tasks.

Technology and Task Complexity: Measured Manning

- **Suitability (of technology):** Technology on the one hand and tasks and functions on the other hand are tuned in to each other.
- **Simplicity (or avoidance of unnecessary complexity):** Training, education, experience, working procedures, *etc.* are tuned in to the complexity of the required tasks.

Questions:

1. How does the organization ensure that the manuals, instructions and operational procedures contain sufficient (as regards different situations, scenarios, *etc.*) and correct (as regards applied technology, *etc.*) information?

   Measures:

   - *Manuals, instructions and operational procedures are implemented based on process documentation.*
   - *A risk-analysis procedure exists for assessing operational tasks (and their sequential order) which need to be carried out.*
   - *Directives and guidance exist concerning the contents of manuals, instructions and operational procedures.*
   - *The company specialists are involved in editing, adapting and customizing the manuals, instructions and operational procedures.*
   - *The users of the manuals, instructions and operational procedures participate in the editing, adapting and customizing process.*
   - *When editing, adapting and customizing the manuals, instructions and operational procedures, a range of circumstances/situations are taken into consideration.*
   - *A formal final inspection and redaction of the manuals, instructions and operational procedures is made by the responsible person at the installation, business unit or company.*

2. How does the organization ensure that the manuals, instructions and operational procedures contain sufficient and correct information in relation to the education or training and the skills of the employee?

   Measures:

   - *The content of the manuals, instructions and operational procedures is compiled according to the function profiles and the actual competencies of the workforce supposed to use them.*

3. How does the organization ensure that the necessary instructions and procedures are present for relevant tasks (or tasks considered to be sufficiently important) within a shift?

   Measures:

   - *Rules and guidelines exist about the necessity to draft instructions and procedures for carrying out routine and non-routine planned tasks (characterized with a certain level of importance and/or complexity).*
   - *It is clearly and unambiguously stated in the instructions and procedures how their users should use them.*

4. How does the organization ensure that the manuals, instructions and operational procedures are easily available to their users?

   Measures:

   - *Manuals, instructions and operational procedures are always promptly and directly available to their users.*
   - *Manuals, instructions and operational procedures are brought together according to a logical format, making easy retrieval of information possible.*
   - *If information is electronically distributed, every user has access to the network.*
   - *If information is electronically distributed, every user is able to make a print-out of the manuals, instructions and operational procedures or he/she has the possibility to let someone print the information for him/her, so that he/she can take a hardcopy of the information to the installation in case this is required to correctly carry out a procedure or instruction.*

5. How does the organization ensure that all necessary explanations, training sessions and education is provided to the users before the implementation of a brand new manual, instruction or operational procedure or before the use of a changed manual, instruction or operational procedure?

   Measures:

   - *All potential users are unambiguously determined for each manual, instruction or operational procedure.*
   - *All potential users are informed about new manuals, instructions or operational procedures and about changes made to manuals, instructions or operational procedures.*
   - *Every modification of a manual, instruction or operational procedure is indicated.*
   - *For the most relevant manuals, instructions and operational procedures, a registration is made of all persons having received an explanation, or having followed an educative course or training.*

- *The responsibilities related to giving courses or trainings are established.*
- *The way in which the explanation, educative course or training is given is defined.*

6. How does the organization ensure that only the most recent versions of the manuals, instructions and operational procedures are in use?

   Measures:

   - *The way in which the organization controls the documents is defined.*
   - *All manuals, instructions, or operational procedures are uniquely identified.*
   - *For every manual, instruction or operational procedure the latest version can easily be identified and retrieved.*

7. How does the organization ensure that the instructions and operational procedures are regularly updated?

   Measures:

   - *The way in which the organization controls the documents, is defined.*

B.3 Procedures and Documentation

Procedures and documentation form a key part within every organization. The importance of procedures and documentation with regard to the operational manning levels lie in following standardized rules for carrying out tasks and lie in elaborating consistent and well-considered documents as regards all responsibilities within the organization. Responsibility schemes, training schedules, competence profiles, absences overviews, reports for smooth transitions from one shift to the next, *etc.*, should be kept up-to-date. The questions thus consider the management, the completeness, and the proper use of procedures and documentation as regards manning levels.

Procedures and Documentation: Ideal Manning

- Manning-level procedures and documentation (inclusive documentation about small changes on manning levels) are well elaborated and are updated on a regular basis and this for various circumstances.
- Procedures and documents are well known and are used properly by all personnel involved.

Procedures and Documentation: Measured Manning

- **Completeness:** All company procedures and documents encompass a variety of situations/circumstances/scenarios.
- **Availability and accessibility:** All company procedures and documents are available and accessible at all times for their intended users.

- **Conformity:** All company procedures and documents match with (i) legislation and company regulations and (ii) user expectations (e.g., taking user training and previous experience into account).

Questions:

1. How does the organization ensure that the documents, procedures, *etc.* concerning the functions, competencies, tasks, trainings, *etc.* are easily available to the users?

   Measures:

   - *The documents, procedures, etc. as regards managing the staffing levels are brought together following a logical format so that information can be easily retrieved.*
   - *If information is electronically distributed, every user has access to the network.*

2. How does the organization ensure that all responsibilities, functions, tasks and the coordination of shift staffing levels are described and known by the employees?

   Measures:

   - *The responsibilities for developing and revising an overview of the functions and responsibilities are fixed.*
   - *The required authorizations for modifying the functions' and responsibilities' overview are defined.*
   - *The responsibility for approving the overview is determined.*
   - *A procedure exists within the organization for communicating the responsibilities to every shift function within a shift and in the case of modifications, for keeping the responsibilities up-to-date.*

3. How does the organization ensure that only the most recent versions of the documents, procedures, *etc.* concerning functions, competencies, tasks, trainings, *etc.* are in use and that these versions are regularly updated?

   Measures:

   - *The way in which the organization controls the documents concerning managing the staffing levels is described.*
   - *All documents, procedures, etc. concerning the management of staffing levels are uniquely identified.*
   - *For every document concerning managing the staffing levels the latest version can easily be identified and retrieved.*

4. How does the organization ensure that the procedures and documents, *etc.* concerning the functions, competencies, tasks, trainings, *etc.* cannot be modified without supervision?

Measures:

- *For every document concerning managing the staffing levels the authorization to perform modifications is unambiguously defined.*
- *When disseminating any modifications made to the documents, procedures, etc. concerning the functions, tasks, competencies, trainings, etc. the necessary explanations about the changes, if required, are given.*

B.4 Communication

Efficient and functional communication between all parties involved in the manning-level-optimization decision process substantially contributes to achieving optimal manning (levels). Periodic exchange of views and (short) meetings in and between company departments may be very important to this end. The interaction conditions as well as the ease to communicate intra-operators and between operators and management also add to optimized manning levels. The questions for this category examine thoroughly the existing communication channels and practices within the organization.

Communication: Ideal Manning

- Communication (sub-organizational as well as organizational) as regards filling in the manning levels is highly adequate and functions independently from the actual manning levels.
- In the case of small manning levels adjustments/changes (e.g., due to optimization of manning levels), all personnel involved (blue collar as well as white collar) are informed.

Communication: Measured Manning

- **Availability:** Communication on filling in the manning levels under various circumstances should be guaranteed within the entire organization and within (and between) parts thereof.
- **Accessibility:** Communication should be accessible and possible between all parties/personnel concerned.
- **Simplicity:** The communication links between all parties should be easy to use and as direct as possible.

Questions:

1. How does the organization ensure that the operational staffing levels and minor modifications thereof are discussed at company, business unit, or installation level and that these discussions could effectively lead to an optimization of the required capacities and skills within the shifts?

Measures:

- *An organizational culture is shaped in which the staffing levels and minor changes thereof are discussed at company, business unit, and installation level.*

2. How does the organization ensure that communication is guaranteed under various circumstances and between all parties concerned (within the installation(s), business unit(s) and the entire organization)?

Measures:

- *The organization provides all necessary communication tools within every installation and business unit of the enterprise, enabling its employees to easily communicate with each other.*
- *When deciding on the means and tools of communication, the potential users are taken into account.*
- *The necessary communication data (phone numbers, etc.) are compiled in an orderly manner and are (with respect to the rules of privacy) easily accessible and available to the potential users.*

3. How does the organization ensure that the communication data and the communication possibilities are periodically revised and updated?

Measures:

- *The communication data made available to potential users are periodically revised.*
- *The organization is responsible for the appropriate functioning of the communication tools.*

B.5 Learning Facilities/Possibilities, Training and Education, Competencies

All management measures taken, all programs and documents with reference to training and education, *etc.* give support to keeping the manning levels of an organization up-to-date and flexible. For example, a well-elaborated manning competence profile is essential for justifiable and solid manning levels. In this category, the questions are characterized by appropriate organization and management of learning facilities and learning possibilities for the company's personnel.

Learning Facilities/Possibilities, Training and Education, Competencies:
Ideal Manning

- The qualities of the functions and people filling in the manning levels are adequately present. Moreover, there is an organizational system in place for further developing the qualities, evaluating/assessing them and–if necessary–adjusting them.

- In the process of determining the qualities required for the functions and people filling in the manning levels, various circumstances are being considered.

Learning Facilities/Possibilities, Training and Education, Competencies: Measured Manning

- **Logic and consistency:** A learning trajectory for every function within operational manning should be established to systematically build up personnel competences profiles and responsibilities (tasks). In this competences build-up process, all shift functions are taken into account and they are treated in a holistic way.
- **Evaluation and remediation:** Competences and training/education of all functions are checked, evaluated and if necessary adjusted at regular time intervals.

Questions:

1. How does the organization ensure that the employees effectively acquire the necessary capacities and skills through the different training and educational facilities/possibilities per function in a shift?

   Measures:

   - *For every function a training and education trajectory is provided.*
   - *Training and education objectives are fixed.*
   - *The contents of all training sessions and educative courses are described.*
   - *The manner in which to give training sessions and educative courses is defined.*
   - *For each training session and educative course the teacher's characteristics are outlined.*
   - *Minimum requirements as regards educational tools are defined.*
   - *For every educative course, the levels to which its objectives are attained are evaluated.*
   - *Every educative course is finished in a formal way.*
   - *The trainings and educative courses are periodically evaluated and revised.*

2. How does the organization ensure that all necessary periodic training and educative courses are given and/or that the necessary evaluations (regarding the competencies to be assumed in reference to the training sessions and to the educative courses) are carried out in an objective way and that they are repeated periodically?

   Measures:

   - *The installation, business unit or organizational needs for training sessions and educative courses are periodically analyzed.*
   - *For certain (pre-defined) training sessions and educative courses, the minimum frequency at which they need to be repeated, is fixed.*

- *For every employee, all his/her followed training sessions and educative courses are registered.*
- *For every training session and educative course, the evaluation specifications are determined (who is entitled to evaluate in which way, etc.).*

3. How does the organization ensure that all the necessary competency profiles of every employee within a shift are present in relation to the total package of required competencies within the shift?

   Measures:

   - *The required minimum competencies for each shift are defined.*
   - *When establishing the trainings and education trajectory for every function in a shift, the competency profiles already present within the shift are taken into consideration.*

4. How does the organization ensure that the competency profiles are kept up-to-date?

   Measures:

   - *The responsibility for developing and maintaining the competency profiles is defined.*
   - *The competency profiles are periodically inspected in a formal way by installation, business unit or company management.*

5. How does the organization ensure that every employee has the necessary maturity and competence to consult on the correct procedures and documents if required?

   Measures:

   - *The maturity and the reflex of being responsible are basic competencies required for all employees within the organization.*

Appendix C Instrument for Evaluating Safety Critical Staffing Levels

1. Supervision / Intervention possibility (Detection)	Yes	No

1.1. Is continuous supervision possible (24/7)

A. in the field (manual interventions)? .... ☐ ☐

B. in the control room (computer-operated interventions)? .... ☐ ☐

C.Q.: In which way can this be guaranteed (qualitatively as well as quantitatively)?

e.g.

- one all-round competent operator is assigned to the control room and to every part of the process installation;
- one operator assigned to the control room supported by several field operators are responsible for certain parts of the process installation;
- several operators are responsible for the control room and for some parts of the process installation (having an all-round competence).

If both answers are rated "yes", ranking 1. =1A, go to 2.1; otherwise go to 1.2.

1.2. Does a discernable and audible alarm exist that is sent to a location (e.g., central CR of plant/cluster) where personnel is permanently (24/7) located should a safety-critical problem arise?

A. in the field (manual interventions)? .... ☐ ☐

B. in the control room (computer-operated interventions)? .... ☐ ☐

C.Q.: Does the possibility exist the alarm might fail?

If both answers are rated "yes", ranking 1. =1B, go to 2.1; otherwise go to 1.3.

Multi-Plant Safety and Security Management in the Chemical and Process Industries. G.L.L. Reniers

ISBN: 978-3-527-32551-1

1.3. Does a back-up system exist to report a safety-critical problem

A. in the field (manual interventions)? .... ☐ ☐

B. in the control room (computer-operated interventions)? .... ☐ ☐

C.Q.: Does the possibility exist the system might fail?

If both answers are rated "yes": ranking 1. =1C; otherwise ranking 1. =1D; go to 2.1.

Extra documents to consult in order to evaluate principle 1: shift team system, operator competence schemes, workschemes, technical information concerning the alarm, technical information concerning the back-up system, *etc.*

---

2. Distractions (Detection)	**Yes**	**No**

2.1. Do the operators have to perform/attend to other specific tasks (e.g., answering phones, administration, production tasks, *etc.*) in addition to safety-critical activities

A. in the field (manual interventions)? .... ☐ ☐

B. in the control room (computer-related interventions)? .... ☐ ☐

C.Q.: If the answer is "no", how can this be guaranteed?

If both answers are rated "no", ranking 2. =2A, go to 3.1; otherwise go to 2.2.

2.2. Does an alternative exist in case the operators miss an alarm due to other tasks

A. in the field (manual interventions)? .... ☐ ☐

B. in the control room (computer-related interventions)? .... ☐ ☐

C.Q.: What is the alternative? (e.g., the alarm is sent to a third party, there is a continuous alarm on the same level, there is a new alarm at the next warning level, *etc.*)

If both answers are rated "yes", go to 2.3; otherwise ranking 2. =2D, go to 3.1.

2.3. Is the time left sufficient to respond to the alarm?

A. in the field (manual interventions)? .... ☐ ☐

B. in the control room (computer-related interventions)? .... ☐ ☐

C.Q.: On which basis is the time margin defined in the industrial area?

If both answers are rated "yes": ranking 2. =2B if time margin exceeds 5 min; ranking 2. =2C if time margin <5 min; otherwise ranking 2. =2D; go to 3.1.

Extra documents to consult in order to evaluate principle 2: listing of (other) operator tasks, technical information concerning the alternative alarm, technical information concerning the time definitions within the company, *etc.*

3. Information (Diagnosis)	Yes	No
3.1. Is it necessary to consult extra information in order to diagnose and solve a possible safety-critical problem		
A. in the field (manual interventions)?	☐	☐
B. in the control room (computer-related interventions)?	☐	☐

C.Q.: If the answer is "no", how will the problem be diagnosed and solved?

If both answers are rated "no", ranking 3. =3A, go to 4.1; otherwise go to 3.2.

	Yes	No
3.2. Is it possible to attain this information 24/7		
A. in the field (manual interventions)?	☐	☐
B. in the control room (computer-related interventions)?	☐	☐

C.Q.: How can the attainability of the information be guaranteed?

Go to 3.3.

	Yes	No
3.3. Is this information (concerning every topic) always correct and intelligible?		
A. in the field (manual interventions)?	☐	☐
B. in the control room (computer-related interventions)?	☐	☐

C.Q.: How can this be guaranteed?

If both questions 3.2 and 3.3 are rated "yes" for all topics, ranking 3. =3B, go to 4.1; otherwise go to 3.4.

	Yes	No
3.4. Does a back-up system exist to provide information in the industrial area		
A. in the field (manual interventions)?	☐	☐
B. in the control room (computer-related interventions)?	☐	☐

C.Q.: Explain the back-up system.

If both answers are rated "yes", ranking 3. =3C, go to 4.1; otherwise go to 3.5.

	Yes	No
3.5. Does another possibility exist in which the problem can be diagnosed and solved		
A. in the field (manual interventions)?	☐	☐
B. in the control room (computer-related interventions)?	☐	☐

C.Q.: Explain the alternative solution. (e.g., cluster know-how with extra information)

If both answers are rated "yes", ranking 3. =3C; otherwise ranking 3. =3D, go to 4.1.

Extra documents to consult in order to evaluate principle 3: listing of safety-critical problems, technique of attainability, technical information about the system which provides back-up information, technical information concerning the alternative solution, etc.

4. Communication links (Diagnosis)	Yes	No
4.1. To diagnose and to solve certain safety-critical problems, is there some kind of communication link required		
A. between the field operators?	☐	☐
B. between the field operator(s) and the control-room operator(s)?	☐	☐

C.Q.: If the answer is "no", how can this be guaranteed?

If both answers are rated "no", ranking 4. =4A, go to 5.1; otherwise go to 4.2.

	Yes	No
4.2. Can this communication link be guaranteed 24/7?		
A. between the field operators?	☐	☐
B. between the field operator(s) and the control-room operator(s)?	☐	☐

C.Q.: How can it be guaranteed?

If both answers are rated "yes", ranking 4. =4B, go to 5.1; otherwise go to 4.3.

4.3. Does a back-up system exist concerning one or more communication links within the plant/cluster which can be guaranteed 24/7 for usage in abnormal situations (NOT emergency situations!)

C.Q.: Explain the back-up system.

If the answer is rated "yes", ranking 4. =4C, otherwise ranking 4. =4D, go to 5.1.

Extra documents to consult in order to evaluate principle 4: listing of communication links, technical specifications concerning the communication links and their assurance, technical information concerning the back-up system, *etc.*

5. Assisting personnel (Recovery)	Yes	No
5.1. To recover certain safety-critical problems, is manning assistance needed		
A. in the field (manual interventions)?	☐	☐
B. in the control room (computer-related interventions)?	☐	☐

C.Q.: If the answer is "no", how can this be guaranteed?

If both answers are rated "no", ranking 5. =5A, go to 6.1; otherwise go to 5.2.

	Yes	No
5.2. Is this manning assistance available 24/7		
A. in the field (manual interventions)?	☐	☐
B. in the control room (computer-related interventions)?	☐	☐

C.Q.: How can this be guaranteed?

Go to 5.3.

5.3. Can this manning assistance arrive in time

A. in the field (manual interventions)? ☐ ☐

B. in the control room (computer-related interventions)? ☐ ☐

C.Q.: How is this time margin defined? Explain.

If both questions 5.2 and 5.3 are rated "yes" for all topics, go to 5.4; otherwise ranking 5. =5D, go to 6.1.

5.4. Does a back-up system exist if the call-up system for summoning the manning assistance fails

A. in the field (manual interventions)? ☐ ☐

B. in the control room (computer-related interventions)? ☐ ☐

C.Q.: Explain this back-up system.

If the answers to questions 5.2, 5.3 and 5.4 are all rated "yes": ranking 5. =5B if time in which availability is ensured is less than 5 min; ranking 5. =5C if time in which availability is ensured exceeds 5 min; otherwise ranking 5. =5D; go to 6.1.

Extra documents to consult in order to evaluate principle 5: shift team system, manning assistance system, technical information about attainability of manning assistance, time-margin calculations, back-up system information, *etc.*

6. Recovery operations (Recovery)	**Yes**	**No**

6.1. Can recovery operations be accomplished within the minimum calculated time margin

A. in the field (manual interventions)? ☐ ☐

B. in the control room (computer-related interventions)? ☐ ☐

C.Q.: How can this be guaranteed? (e.g., task analyses, desktop exercices, simulations, *etc.*)

If both answers are rated "yes", go to 6.2.; otherwise go to 6.5.

6.2. Is it possible that the operators

A. are assigned extra recovery operations (e.g.: back-up activities of other operators within the plant/cluster)? ☐ ☐

B. are assigned extra tasks (e.g.: responsibility for site alarm, emergency phone services, responsibility for non-critical alarms? ☐ ☐

C.Q.: If the answer is "no", how can this be guaranteed?

If both answers are rated "no", ranking 6. =6A, stop audit; otherwise go to 6.3.

6.3. Are they timely informed about the extra assignments?

C.Q.: How can this be guaranteed?

If the answer is rated "yes", go to 6.4; otherwise ranking 6. =6D, stop audit.

6.4. Can all assigned tasks be accomplished (originally assigned recovery operations, extra assigned recovery operations and extra tasks)

A. by field operators? .... ☐ ☐

B. by control-room operators? .... ☐ ☐

C.Q.: How can this be guaranteed?

If both answers are rated "yes", ranking 6. =6B, stop audit; otherwise go to 6.5.

6.5. Does a back-up solution (at plant/cluster level) exist should it prove impossible to accomplish the tasks

A. for field operators? .... ☐ ☐

B. for control-room operators? .... ☐ ☐

C.Q.: Explain the back-up system. (e.g., extra back-up operator, assistance personnel to handle the administrative tasks and technical support)

If both answers are rated "yes", ranking 6. =6C; otherwise ranking 6. =6D; stop audit.

Extra documents to consult in order to evaluate principle 6: recovery time-margin calculations/simulations, technical recovery information/data, back-up system data, *etc.*

Index

In this index tables are indicated in italics; figures in bold and appendices by a.

Multi-Plant Safety and Security Management in the Chemical and Process Industries. G.L.L. Reniers

ISBN: 978-3-527-32551-1